# SpringerBriefs in Education

We are delighted to announce SpringerBriefs in Education, an innovative product type that combines elements of both journals and books. Briefs present concise summaries of cutting-edge research and practical applications in education. Featuring compact volumes of 50 to 125 pages, the SpringerBriefs in Education allow authors to present their ideas and readers to absorb them with a minimal time investment. Briefs are published as part of Springer's eBook Collection. In addition, Briefs are available for individual print and electronic purchase.

SpringerBriefs in Education cover a broad range of educational fields such as: Science Education, Higher Education, Educational Psychology, Assessment & Evaluation, Language Education, Mathematics Education, Educational Technology, Medical Education and Educational Policy.

SpringerBriefs typically offer an outlet for:

- An introduction to a (sub)field in education summarizing and giving an overview of theories, issues, core concepts and/or key literature in a particular field
- A timely report of state-of-the art analytical techniques and instruments in the field of educational research
- A presentation of core educational concepts
- An overview of a testing and evaluation method
- A snapshot of a hot or emerging topic or policy change
- An in-depth case study
- A literature review
- A report/review study of a survey
- An elaborated thesis

Both solicited and unsolicited manuscripts are considered for publication in the SpringerBriefs in Education series. Potential authors are warmly invited to complete and submit the Briefs Author Proposal form. All projects will be submitted to editorial review by editorial advisors.

SpringerBriefs are characterized by expedited production schedules with the aim for publication 8 to 12 weeks after acceptance and fast, global electronic dissemination through our online platform SpringerLink. The standard concise author contracts guarantee that:

- an individual ISBN is assigned to each manuscript
- each manuscript is copyrighted in the name of the author
- the author retains the right to post the pre-publication version on his/her website or that of his/her institution

More information about this series at http://www.springer.com/series/8914

Elaine Khoo · Craig Hight · Rob Torrens
Bronwen Cowie

# Software Literacy

## Education and Beyond

 Springer

Elaine Khoo
Faculty of Education, Wilf Malcolm Institute
  of Educational Research (WMIER)
University of Waikato
Hamilton, Waikato
New Zealand

Rob Torrens
Faculty of Science and Engineering,
  School of Engineering
University of Waikato
Hamilton, Waikato
New Zealand

Craig Hight
School of Creative Industries
The University of Newcastle
Newcastle, NSW
Australia

Bronwen Cowie
Faculty of Education, Wilf Malcolm Institute
  of Educational Research (WMIER)
University of Waikato
Hamilton, Waikato
New Zealand

ISSN 2211-1921         ISSN 2211-193X   (electronic)
SpringerBriefs in Education
ISBN 978-981-10-7058-7         ISBN 978-981-10-7059-4   (eBook)
https://doi.org/10.1007/978-981-10-7059-4

Library of Congress Control Number: 2017956740

Printed on acid-free paper

This Springer imprint is published by Springer Nature
The registered company is Springer Nature Singapore Pte Ltd.
The registered company address is: 152 Beach Road, #21-01/04 Gateway East, Singapore 189721, Singapore

# Preface

Inspired by the emerging field of Software Studies (Fuller, 2008; Kitchin & Dodge, 2011; Manovich, 2013), this book aims to introduce the notion of 'software literacy' as an emerging area of research and practice for educators, researchers and policymakers. As a cultural artefact, software plays a role in reproducing, reinforcing and augmenting existing cultural practices, as well as generating new practices (Manovich, 2013). Software platforms, such as Facebook and YouTube, operating systems such as iOS software in iPhones and iPads and everyday applications such as the Microsoft Office suite are just some examples of how software has become embedded in everyday personal and professional pursuits. The role that software plays in our lives is however largely unacknowledged and invisible. The notion of software literacy directs attention to this influence. Developing a critical software literacy is, we argue, an essential part of learning and living in the twenty-first century, and a capacity that transcends the use of any particular software tool and any particular educational, social and cultural context.

Although our focus is primarily an educational one, our argument has implications for any field that makes use of software and information and communication technology (ICT) systems and applications. The book will be of interest to those engaging with the challenges and opportunities involved in software-based teaching and learning, and to people who are interested in how software impacts on the workplace and leisure activities that are part of our day-to-day lives.

The book sets out findings from our two-year university-level study, which is one of the very few studies that have investigated the nature and development of software literacy. Specifically, it provides case studies of software use and literacy demands in university-level engineering, and screen and media studies courses. The views of lecturers and students are presented—student views are mapped over their full programme of study to illustrate a possible framework for the development of software literacy.

The cases provide a forum for our elaboration of the tensions and productive exchanges between the pragmatic and creative potential of software use. They illuminate how the nature and use of software are entangled with the history of the two disciplines. Just as importantly, they illustrate how software shapes and is part of student opportunities to learn and the learning networks that span formal and informal activities inside and outside the 'classroom'.

The book concludes by asserting and scoping the need for the general population to be software literate and for professionals within software-based disciplines to be critically aware of the way their choice of software enables and constrains their actions and informs their creative imagination.

| | |
|---|---|
| Hamilton, New Zealand | Elaine Khoo |
| Newcastle, Australia | Craig Hight |
| Hamilton, New Zealand | Rob Torrens |
| Hamilton, New Zealand | Bronwen Cowie |

## References

Fuller, M. (2008). *Software Studies: A Lexicon*. Cambridge, MA: The MIT Press.
Kitchin, R. & Dodge, M. (2011). *Code/Space: Software and Everyday Life*. Cambridge, MA: The MIT Press.
Manovich, L. (2013). *Software Takes Command* (International Texts in Critical Media Aesthetics, Vol. 5). NY: Bloomsbury Press.

# Acknowledgements

The authors gratefully acknowledge funding support from the Teaching and Learning Research Initiative, New Zealand Council for Educational Research, Wellington, New Zealand.

We also thank all the participants who took the time and made the effort to take part in the study. The contribution and insights you have provided have been invaluable to this research and this book.

We also recognise and thank Gareth Ranger for his contribution to this research project during his time as a University of Waikato Summer Research Scholar.

# Contents

# List of Figures

# Chapter 1
# Introduction: Software and Other Literacies

**Abstract** This chapter outlines the role and significance of software in contemporary society. Drawing from the new field of Software Studies, it sets outs key concepts relevant to the study of software, including affordances, agency, human-machine assemblages, and performance to explain the ways users co-create with software. It proposes the notion of *software literacy* as a framework to help readers unpack the ways the affordances of software can (re)shape the ways we think and act. These ideas are then grounded in an examination of an educational research project into the ways in which students become more *literate* about the nature and implications of software which they encounter as part of their tertiary studies.

## 1.1 Introduction

This book addresses a question which we argue needs to be answered in detail and for a wide variety of contexts: why and how does software matter? The research presented in this volume responds to the more general prompt of Software Studies, a comparatively new field of inquiry that Lev Manovich and others have championed (Fuller, 2003, 2008; Hawk, Rieder, & Oviedo, 2008; Johnson, 1997; Kitchin & Dodge, 2011; Manovich, 2013). This is a paradigm which derives from an insistence that software has become the engine of contemporary information society as much as electricity and the combustion engine made the industrial revolution possible (Manovich, 2013); code is now part of the infrastructure of modern societies. Further, we are living in a software culture, one which is fundamentally reshaping all areas of modern life:

> I think of software as a *layer that permeates all areas of contemporary societies.* Therefore, if we want to understand contemporary techniques of control, communication, representation, simulation, analysis, decision-making, memory, vision, writing, and interaction, our analysis can't be complete until we consider this software layer. Which means that all disciplines which deal with contemporary society and culture—architecture, design, art criticism, sociology, political science, humanities, science and technology studies, and so on—need to account for the role of software and its effects in whatever subjects they investigate (Manovich, 2008, p. 8, emphasis in original).

© The Author(s) 2017
E. Khoo et al., *Software Literacy*, SpringerBriefs in Education,
https://doi.org/10.1007/978-981-10-7059-4_1

This is a startling and provocative series of claims, but nevertheless a useful one to consider. It is not hard to find examples from everyday life where we engage with software in various forms: when we are playing gaming apps on our smartphones, when we are accessing banking systems at an ATM, when we are using recommendation systems to find material on Netflix or YouTube, when using everyday social media platforms such as Facebook, Twitter and SnapChat, which provide a broader infrastructure of engagement, or wayfinding in a strange landscape by using a navigation system in our car, or using various apps to track our individual biometric measures as part of a general effort to improve our health. Despite the diversity of hardware involved, at base these all involve us as users engaging with different kinds of applications, platforms and infrastructures constituted through software code. In many cases they establish an apparent set of *givens* for how users, practitioners and citizens can engage, participate and interact (van Dijck, 2013). All contemporary media practices, and increasingly a range of other social, political and economic practices, are now clearly embedded within and deeply informed by their nature as software. At an infrastructural level, for example, the internet and World Wide Web (Web hereon) are themselves organised through software-based protocols (Galloway, 2004) that govern largely automated processes that are rarely visible to everyday users, unless they fail. Once we peel back the layers of contemporary society, we recognise that software also runs in the background of many of our key institutions and systems, from the information systems of a hospital, the planning and organisation of schools, the disturbingly sophisticated communication and targeting capabilities of the military-industrial complex, to the automated financial exchanges that drive global share markets. It is perhaps useful to think of these kinds of everyday practices as *coded*, in the sense that they are deeply embedded within and enabled computer code; they have been translated into software and augmented or transformed into something else because of this fact. And, crucially, these are often now practices which no longer operate or exist outside of programming code.

Manovich (2013) argues that this has profound implications for understanding aspects of culture such as the emergence of the internet and Web in the 1990s. The Web is not just a new means for distribution, but also constitutes a medium itself with distinctive characteristics, and underlies more recent and interconnected developments within mobile and gaming devices—not least the emergence of entire software-based ecosystems such as Apple's iOS and the parallel Android mobile ecosystem. These are all developments which constitute a rapidly expanding universe of software culture, fed by recombinatory, evolutionary growth—as a capability, or function or set of tools become coded they become available to be recombined in new ways for different platforms and contexts. As new spheres of human activity become coded, they become part of a broader emergence and dominance of software culture.

In the earlier quote above, Manovich is dealing with abstract generalisations, but as our examples illustrate, software needs to be acknowledged as the dominant cultural technology of our time at multiple levels. Such totalising statements do not provide sufficient nuance to address the variety of social-cultural, economic and political contexts in which cultural technologies are embedded, nor do they provide the nuance of insight for us to fully consider the implications of software for the way we lead our lives and the aspirations we have and pursue. Without investigation into micro-contexts, we run the danger of technological determinism and utopian or apocalyptic rhetoric familiar to new technologies (and which educational institutions are particularly prone to). Our research is one small response to both the wider provocations of Software Studies, and a necessary empirical corrective to its broader claims.

## 1.2 Software Studies and Its Significance Within Contemporary Society

Software Studies adopts the perspective that the study of software partly involves investigating the cultural discourses that are embedded in code, together with the broader implications for users of how these discourses operate through the application of that code. Coding (or programming, as these terms tend to be used interchangeably) is a form of writing which inscribes types of actions to be performed using a computer.[1] To some extent software is a neglected part of the digital revolution. Software is not always thought of as cultural product, arising from specific contexts and bearing the legacy of the institutions, practices and personnel who created it. Software constitutes a large industry that is not always theorised, considered and investigated within media, education and other parts of society. *Software* was not recognised as a distinct industry until the 1960s, developing in the wake of the emergence of the personal computer age in the 1970s (driven especially by the release of Apple II in 1977) and now operating across diverse spheres. Practitioners of this industry now describe themselves as *software engineers*:

> Software engineering describes software development as an advanced writing technique that translates a text or a group of texts written in natural language (namely, the requirements specifications of the software 'system') into a binary texts or group of texts (the executable computer programs), through a step-by-step-process of gradual refinement (Frabetti, 2014, p. xx).

The term *engineer* here usefully obscures one of the key characteristics of software; that it is a form of writing that not only has material effects but also often has unintended consequences. No form of code is perfect; it emerges from human endeavour and is inscribed with the conditions of its creation as with all cultural artefacts. Software is also an evolving part of culture, "an essentially unfinished product, a continually updated, edited and reconstructed piece of machinery" (Berry, 2011, p. 39), with components that may have their own life cycle, break down or be recombined towards new ends (Berry, 2011).

Crucially, many software applications also foster other creative acts, as Kitchin and Dodge outline, "software is itself a medium for intellectual work and invention" (Kitchin & Dodge, 2011, p. 112).

---

[1]In more technical terms, a programmer writes in a language (source code) which is then translated, or compiled, into another language (object code) the computer can parse.

For us, software needs to be theorized as both a contingent product of the world and a relational producer of the world. Software is written by programmers, individually and in teams, within diverse social, political, and economic contexts. The production of software unfolds – programming is performative and negotiated and code is mutable. Software possesses secondary agency that engenders it with high technicity. As such, software needs to be understood as an actant in the world – it augments, supplements, mediates, and regulates our lives and opens up new possibilities – but not in a deterministic way. Rather, software is afforded power by a network of contingencies that allows it to do work in the world. Software transforms and reconfigures the world in relation to its own systems of thought (Kitchin & Dodge, 2011, pp. 43–44).

The notion of software entailing forms of agency is an important insight here. As we come to rely increasingly on software-embodied practices, as more parts of society become *coded*, in a broader sense there are necessary debates over the extent to which we are collectively as a species ceding creative, conceptual and communicative agency to platforms and infrastructures, and hence key parts of how we imagine the possibilities for those practices in the first place.

## 1.3  Cultural Software and Agency

Within the Software Studies paradigm, Manovich identifies *cultural software* as that which is central to cultural production, in its broadest terms. Cultural software includes the popular and ubiquitous forms of software that we use ourselves in word processing, image manipulation, and even in gaining access to cultural works such as through internet browsers and media players.

I am using this term literally to refer to *software programs which are used to create and access media objects and environments*. The examples are programs such as Word, Power-Point, Photoshop Illustrator, Final Cut, After Effects, Flash, Firefox, Internet Explorer, etc. Cultural software, in other words, is *a subset of application software* which enables creation publishing, accessing, sharing, and remixing images, moving image sequences, 3D designs, text, maps, interactive elements, as well as various combinations of these elements such as web sites, 2D designs, motion graphics, video games, commercial and artistic interactive installations, etc. (Manovich, 2008, p. 13, emphasis in original).

These kinds of software collectively foster a wide range of activities, recognisable to most readers of this volume; everything from creating and sharing forms of media content such as photos or videos, playing computer games, posting content on Wikipedia or Facebook, communicating with other users through texting, email or other services, operating a search engine, browsing on the Web, and so on (Manovich, 2011). This is a list which touches on all forms of computer-based and especially networked media, and involves a wide variety of software targeted at both everyday novices and more professional users. (Of this broader category of cultural software, we are looking in the research presented in this volume at a small sample of desktop applications deployed in educational contexts, such as the Adobe Creative Suite of media editors, and Computer-Aided Design software.)

At the level we as users encounter an application or platform, our engagement is both fostered and constrained through the affordances which that piece of software provides.

> An affordance is an action possibility or an offering. Possible actions on a computer system include physical interaction with devices such as the screen, keyboard, and mouse [...] The application software also provides possible actions. A word processor affords writing and editing at a high level, but it also affords clicking, scrolling, dragging and dropping. The functions that are evocable by the user are the affordances in software. Functions may include text-editing, searching, or drawing. The information that specifies these functions may be graphical (buttons, menus) or it may not exist at all (McGrenere & Ho, 2000, p. 6).

Affordances allow us to do particular things; to select, to view, to manipulate in specific ways. If we look at a software application as providing a set of these possible actions, then it is vital to map how these affordances appear within a specific hierarchy, with some made easily available to its users, and how they are more generally organised to support or constrain what users can use that application for.

The interface for a piece of software embodies that hierarchy of affordances; these are the default tools we find most easily on the ribbons or drop-down menus of one of the Microsoft Office applications, or the buttons that are clustered together on a social media platform (such as Facebook's now expanded range of feedback buttons). At a more fundamental level, if we extrapolate from the set of affordances which a piece of software provides, we can start to see the underlying *conceptual* framework which an application or platform operates within. As particular affordances become familiar to users, and become naturalised to some extent within specific forms of practice, they can become associated with specific ways of thinking. When we look at a specific application, we need to analyse the ways in which it encourages particular ways of thinking and working through creative and cultural practices—the manner in which its design is informing and shaping how we imagine the creative possibilities for action using that application.

There are key aspects of software culture that are useful to consider when seeking to understand the contention that software entails a form of agency. A core premise of Software Studies is the need to move away from seeing software platforms and applications as neutral, as simply things that you do something with (Fuller, 2003, p. 16). Programming code needs to be understood broadly as engendering *both forces of empowerment and discipline*. Software applications and platforms are attractive precisely because they are designed toward increasing efficiencies and productivity, generating entirely new markets, and providing new forms of play and creativity. However, they also serve as "a broad range of technologies that more efficiently and successfully represent, collate, sort, categorize, match, profiles, and regulate people, processes, and places" (Kitchin & Dodge, 2011, pp. 10–11). This tension between *empowerment* and *discipline* offers a broad frame for understanding the layered and complex role which software plays at a variety of levels, especially within networked media. At the more micro level, we need to be considering the manner and ways in which specific pieces of software work to both enable and constrain creative practices.

We argue that human agency operates in a complex way within software culture; we become part of human-machine assemblages where agency becomes more contingent on a range of human and non-human factors. The broader plateau of software culture has more profound implications than just facilitating the creation of digital content through desktop applications:

> (1) in a way that is completely new, software allows the delegation of mental processes of high sophistication into computational systems. This instils a greater degree of agency into technical devices than could have been possible within mechanical systems; (2) networked software, in particular, encourages a communicative environment of rapidly changing feedback mechanisms that tie humans and non-humans together into new aggregates. These then perform tasks, undertake incredible calculative feats, and mobilise and develop ideas at a much higher intensity than in a non-networked environment; (3) there is a greater use of embedded and quasi-visible technologies, leading to a rapid growth in the quantification that is taking place in society (Berry, 2011, p. 2).

This provides the broader context for analysing the specific forms of software which are the key focus of this book, the set of applications used as creative tools within specific disciplines.

## 1.4  Co-creating with Software Tools

In very fundamental ways, it is important to recognise that we *co-create* with cultural software, exploring and negotiating their potential to enable and constrain specific practices. There are implications here for our own creative agency, not just in the more specific affordances provided by applications, but in a broader sense with the imaginative possibilities which software draws from and in turn helps to inform and shape.

> Contemporary media is experienced, created, edited, remixed, organized and shared with software. [...] To understand media today we need to *understand media software* – its genealogy (where it comes from), its anatomy (interfaces and operations), and its practical and theoretical effects. How does media authoring software shape the media being created, making some design choices seem natural and easy to execute while hiding other design possibilities? How does media viewing/managing/remixing software affect our experience of media and the actions we perform on it? How does software change what "media" is conceptually? (Manovich, 2011, para 1).

In large part, we are talking in this book about software which waits to be *performed*, which needs to be interacted with in order to function and realise creative possibilities.

Software Studies as a field is currently focused more on analysing the logics embedded within software applications themselves than building convincing models of how specific groups of users understand and use the affordances of particular software tools, within specific contexts (Hight, 2015). However, it is obviously crucial that we recognise that all forms of software need to be initiated, to be *run*, in order to function. Programming code only becomes of interest when it is put into action by users (or perhaps by other software which has been initiated by users elsewhere).

We characterise users' encounters with software applications using a particular understanding of the notion of *performance*. Drawing, in particular, from Brenda Laurel's work (1992), Manovich uses the term more generally to describe all of the ways in which we interact in professional or more everyday settings with software (Manovich, 2008). At a fundamental level, we are collaborating with programming code when we engage with, respond to, or create content using an application or platform. But this is also the point where the *empowering* or *disciplining* possibilities of software are actualised. When using the most basic operations of a word processor, for example, nothing happens without the active intervention of the user. Generating textual content, as with any creative work involving software, involves human users in collaborative performances with a machine. A blank page in a word processor application is in one sense a large set of creative possibilities. However, the application also obviously informs and shapes what and how we perform options for navigating, editing and sharing of our content. In this sense, there is assumed to be a complex interplay between affordance and performance, which potentially plays out in a unique way each and every time a user engages with any application.

There is an increasing imperative to consider the nature of user performance of software because so much of contemporary media and other practices entail software performances in real-time, creating experiences that only exist at the product of interaction.

> In software culture, we no longer have "documents", "works", "messages" or "recordings" in 20th century terms. Instead of fixed documents whose contents and meaning could be determined by examining their structure and content (a typical move of the 20th century cultural analysis and theory, from Russian Formalism to Literary Darwinism) we now interact with dynamic "software performances." I use the word "performance" because what we are experiencing is constructed by software in real time. So whether we are exploring a dynamic web site, play a video game, or use an app on a mobile phone to locate particular places or friends nearby, we are engaging not with pre-defined static documents but with the dynamic outputs of a real-time computation happening on our device and/or the server. [...] Therefore, although some static documents may be involved, the final *media experience constructed by software usually does not correspond to any single static document stored in some media* (Manovich, 2012, para 7, emphasis in original).

These are broad generalisations, but once again they serve as a prompt to consider in more specific detail how particular pieces of software are used in specific contexts for specific outcomes or purposes.

Most software have a genealogy we can trace, to consider the assumptions about users, and about specific practices, which are *coded* into the possible actions it enables and constrains. In a broader sense, however, it is also important to remember that software evolves in tandem with specific user communities; typically it is early adopters who help to reinforce, refine or reject particular toolsets through their use. There are any number of everyday examples which could illustrate this. Word processing is an early example of application software which became ubiquitous in early personal computers and helped to define what these machines might allow everyday users to achieve. There is comparatively little research, however, which engages with Word processing as software, as a new kind of coded practice. An implication of Word and other early software-based media editors was their fostering of a *cut, copy and paste* sensibility toward creating media content (Fuller, 2003; Heim, 1987; Johnson, 1997; Kirschenbaum, 2016).

> As an example of how the interface imposes its own logic on media, consider "cut and paste" operations, standard in all software running under the modern GUI [Graphic User Interface]. This operation renders insignificant the traditional distinction between spatial and temporal media, since the user can cut and paste parts of images, regions of space, and parts of a temporal composition in exactly the same way. It is also "blind" to traditional distinctions in scale: the user can cut and paste a single pixel, an image, or a whole digital movie in the same way. And last, this operation also renders insignificant the traditional distinctions between media: "cut and paste" can be applied to texts, still and moving images, sounds, and 3-D objects in the same way' (Manovich, 2001, p. 65).

Cut, copy and paste, then, has served as a more profound cultural logic, as these are affordances now naturalised as part of the wider development of graphic user interfaces (GUIs) on personal computers, legitimising and embedding this logic across a variety of software. The diffusion of such tools coincided with the growth of the Web, which allowed for easy exchange, sharing and recycling of different kinds of digital material (Manovich, 2001, p. 131). And, as Manovich notes, this embedding of cut, copy and paste across a variety of media editors has potential implications for how users *conceive* of the possibilities of creating content, and perhaps even for conceptualising what *media* are (Manovich, 2013). Such broader implications of software are aspects of cultural history that we can only address through detailed longitudinal studies—the work outlined in the following chapters (see Chaps. 3 and 4) is one small step in this direction.

## 1.5  Software Literacy: Our Framework for Interrogating Cultural Software

At the core of our research into the implications of software within educational micro contexts is a set of assumptions underlying the notion of *software literacy*. We view software literacy as an integral and fundamental part of the broader label of *digital literacy*, although the current debates around this do not acknowledge software as an actant in the world (Kitchin & Dodge, 2011). Our arguments here are intended to contribute to existing bodies of work on digital literacy (see for example, Ala-Mutka, 2011; Alexander, Adams Becker, Cummins, & Hall Giesinger, 2017; Sefton-Green, Nixon, & Erstad, 2009) and specifically to inform and prompt greater consideration of the importance of *code* as a factor in individual action and an infrastructural part of contemporary society.

Vee articulates a necessarily broad definition of literacy itself as "a widely held, socially useful and valued set of practices with infrastructural communication technologies" (Vee, 2017, p. 216). A plethora of definitions of digital literacy have been put forward, see for example Beetham and Sharpe (2010), JISC (2014), and UNESCO (2013). One of the most comprehensive definitions, according to Ala-Mutka (2011), is offered by Martin and Grudziecki (2006, p. 255), that is:

> Digital Literacy is the awareness, attitude and ability of individuals to appropriately use digital tools and facilities to identify, access, manage, integrate, evaluate, analyze and synthesize digital resources, construct new knowledge, create media expressions, and communicate with others, in the context of specific life situations, in order to enable constructive social action; and to reflect upon this process.

Martin (2008), as Ala-Mutka note, further elaborates the importance of digital actions within everyday situations meaning that what constitutes digital literacy will vary across individuals, contexts and time, and that digital literacy is multidimensional, involving evaluation of and reflection on digital action and development. The different definitions place varying emphases on the cognitive, cultural, socio-emotional and technical dimensions of digital literacy but on the whole they emphasise digital literacy as *user skills*.

Livingstone (2004) offers a contrasting view. Speaking about media literacy, which is widely seen as part of digital literacy, she argues that it is defined through people's "relations with different media rather than defined independently of them" (p. 8, see also Tsatsou, 2017). Seen this way, digital literacy is conceptualised as *user-technology interactivity*. Livingstone emphasises that user-technology interactivity underpins how users interpret, diverge from, conform to or recreate meanings in the process of engaging with digital media (Livingstone, 2008). Livingstone's conceptualisation suggests that we need to consider the interactivity between the user and technology for literacy development, with this interactivity understood as two-way. The idea of a plurality of media literacies challenges the prevalent perception of literacy as a static realm of skills, ability and expertise that determines experiences with technology before use (also see for example Goodfellow, 2011 and a new book series by Lankshear, Knobel, & Peters, 2016). The ability of access, use, act with, evaluate and reflect on digital tools and facilities are all types of social practice that are embedded within and shaped by programming code, and are evolving in tandem with specific sets of software tools.

Drawing from a software studies paradigm, Vee talks of coding as a form of literacy which underpins other activities and knowledges but when we look across these various definitions and bodies of work on digital literacy the role of software itself in shaping action tends to be taken for granted and is not questioned. Our goal in putting forward the notion of software literacy is to highlight the interactivity between users and technologies and then to go further and consider the forms of agency that technologies exert in mediating and shaping action.

Software is not neutral. Software needs to be viewed as increasingly underpinning all aspects of how we operate as *creative producers* (Alexander et al., 2017). The fact that software is part of the infrastructure of our everyday lives, however, does not mean that we inevitably develop critical literacies toward its nature, significance and implications for our lives (also see for example Goodfellow & Lea, 2014). In a now-dated set of debates, numerous authors argued that ubiquitous access to digital technologies informed a new *net* generation of digital natives (Oblinger, 2003) with the corresponding assumption that access to digital tools had, *on their own*, facilitated the development of new learning skill sets (Tapscott, 2009). Educators thus often worked with the assumptions that students already possessed the necessary computing skills and conceptual frameworks to learn with and through generic software packages (Bennett, & Maton, 2010). They consequently tended to neglect the agency of the software itself in shaping how students performed the software (Adams, 2006). More recent research indicates that such assumptions about students' digital proficiencies are unfounded and that digital inequalities and marginalisation persist around students' access to and use of information and knowledge (Bennett, Maton, & Kervin, 2008; Kennedy, Judd, Churchward, Gray, & Krause, 2008). Digital inequality is not restricted to the issue of physical access to software and hardware. Inequalities also arise due to differences in the social and cultural support for ICT use; individuals may be more or less able to perform and critique the affordances software and hardware (Selwyn & Facer, 2007).

We propose the concept of *software literacy* as the repertoires of skills and understandings needed for students to be critical and creative users of software applications and systems in a software saturated culture. This is a framework developed from experience in teaching software, the findings from a number of eLearning research projects, but also informed by (critical) readings of current digital literacy frameworks which, we argue, insufficiently acknowledge the significance of software itself within the ecology of digital literacies. Software literacy, then, inevitably interacts with other literacies under the 'digital' umbrella; how users understand and perform software tools, such as desktop applications, digital platforms that aim to foster a range of creative and communicative practices, depends on the host of understandings, skills and competencies and experiences that users bring to the moment of engagement.

We hypothesise that there are three progressive tiers of development towards software literacy:

1. a foundational skill level where a learner can use a particular software, recognising and deploying its key affordances, but would, for example, be more likely to rely on default settings and affordances rather than confidently customising the software's interface,
2. an ability to independently troubleshoot and problem-solve issues faced when using the software, reaching a key threshold where users are more likely to independently draw upon a variety of resources in their troubleshooting (i.e. being able to facilitate their own learning of the software when encountering affordances they do not know),
3. and finally, the ability to critique the software, including being able to apply such critique to a range of software designed for a similar purpose and to use these understandings for new software learning. The third tier involves the ability to identify affordances and their implications (including the constraints) of particular software and identify ways to both apply and extend its use such that it is relevant and meaningful to a wider range of learning purposes, tasks and contexts. In learning environments this is often the point, for example, where students feel they have *mastered* at least some of a particular software application and can apply their skills to a variety of purposes (including those not necessarily anticipated by the application's designers). Students have encountered similar or complementary

tools, and start to appreciate what each application is (and is not) most useful for, and develop the ability to *talk back* to the software; to critique its limitations and engage in debates over how it *should have been* designed.

Our own experience has demonstrated that incoming students to our tertiary institutions may not be aware of the full implications of the affordances and constraints offered through particular software. No studies to date that we know of raise the role of student understanding of how software and its affordances influence knowledge representation, generation and critique. We know very little about how students develop the skills and expertise needed to attend to the features and use of software (as application, platform and architecture) to complete everyday tasks. There is evidence that the ubiquity of software and ICT tools has led students to adopt a range of informal approaches to meet their learning needs (Khoo & Cowie, in press; Peeters et al., 2014). Research also indicates that students' formal software learning backgrounds are diverse (Khoo, Johnson, Torrens, & Fulton, 2011). Student knowledge and use of software and technology is highly specific to their formal and informal educational, social and cultural contexts for learning and use (Jones, Ramanau, Cross, & Healing, 2010; Valtonen, Dillon, Hacklin, & Väisänen, 2010). The challenge is thus posed for educators to adopt pedagogical strategies that build on this diversity. Our research (see Chaps. 3 and 4), small and necessarily tightly focused as it is, is intended to demonstrate the value of detailed empirical work in this area of educational research (and more broadly the need for investigation into how users acquire the skills and understandings which inform and shape their ability to co-create with various forms of software).

Initial baseline studies of our work exploring tertiary student software literacies associated with Microsoft PowerPoint as a common software package encountered and used in teaching and learning at the tertiary level indicated that despite a familiarity with PowerPoint and its many affordances, majority of students were not able to critique the ways the software shaped their understanding of disciplinary knowledge (i.e. they showed a relative lack of critique at the third level of our software literacy framework) (Khoo, Hight, Cowie, Torrens, & Ferrarelli, 2014). In this volume, we draw on findings from an extension of the baseline study to focus on two case studies of teaching and learning contexts with discipline-specific software learning in Media Studies (e.g., Adobe Photoshop, Final Cut Pro), and Engineering (SolidWorks) to explore the effect of more direct instruction on student software literacy development and disciplinary understanding.

## 1.6   Moving Forward in This Book

With this chapter laying the foundations of our thinking regarding the important role of software and our proposed notion of software literacy, Chap. 2 goes further to elaborate on these ideas by visiting the genealogy and development of software in two fields—media studies and engineering, which make use of Digital Non-Linear

Editing (DNLE) and Computer-Aided Design (CAD) software respectively—and its impact on the small set of cultural and creative practices which are the focus of our own research. Based on developments in Chap. 2, in Chaps. 3 and 4, we describe an educational research project aimed at understanding university students' software literacy development across two diverse disciplines of study (media studies and engineering) at one New Zealand institution. These represent disciplines where most students are expected to develop some understanding or even proficiency in complex software platforms. The research is important in its investigation of the extent students develop the knowledge and skills needed for them to use and critique discipline-specific software. We explored students' wider practices in learning software (formally and informally); how their software literacies were nurtured (or not) within specific learning contexts (university, and workplace in the case of engineering students). The media studies and engineering cases provide usefully overlapping but also contrasting examples of approaches toward the teaching of software and students' software learning trajectories. Chapter 5 provides a comparative analysis of the two case studies reported in Chaps. 3 and 4 to highlight similarities and differences in student experiences with learning and using discipline-specific software. Common themes shared by both cases are highlighted and differences noted in relation to our software literacy framework. We provide further discussions regarding the design, support, teaching and learning of students where software plays a central role in the understanding and application of disciplinary ideas in tertiary contexts.

Finally, Chap. 6 pulls together the key themes from the book, revisits our arguments for the notion of software literacy as an essential part of learning and living in the 21st century. Put another way, we assert that being aware of the influence of software itself is central to the digital literacy needed for critical and creative participation in society and work today. Chapter 6 looks to prompt further research activity in educational and related research fields.

# References

Adams, C. (2006). PowerPoint, habits of mind, and classroom culture. *Journal of Curriculum Studies, 38*(4), 389–411.

Ala-Mutka, K. (2011). Mapping digital competence: Towards a conceptual understanding. Seville: JRC-IPTS. Retrieved from http://ipts.jrc.ec.europa.eu/publications/pub.cfm?id=4699.

Alexander, B., Adams Becker, S., Cummins, M., & Hall Giesinger, C. (2017). Digital literacy in higher education, Part II: An NMC horizon project strategic brief (Volume 3.4). Austin, Texas: The New Media Consortium. Retrieved from https://blog.stcloudstate.edu/ims/files/2017/08/2017-nmc-strategic-brief-digital-literacy-in-higher-education-II-ycykt3.pdf.

Beetham, H., & Sharpe, R. (2010). Digital literacy framework. Retrieved from http://jiscdesignstudio.pbworks.com/w/page/46740204/Digital%20literacy%20framework.

Bennett, S., & Maton, K. (2010). Beyond the 'digital natives' debate: Towards a more nuanced understanding of students' technology experiences. *Journal of Computer Assisted Learning, 26*(5), 321–331. doi:10.1111/j.1365-2729.2010.00360.x.

Bennett, S., Maton, K., & Kervin, L. (2008). The 'digital natives' debate: A critical review of the evidence. *British Journal of Educational Technology, 39*(5), 775–786.

Berry, D. M. (2011). *The philosophy of software: Code and mediation in the digital age*. Houndmills, United Kingdom: Palgrave Macmillan.

Frabetti, F. (2014). *Software theory: A cultural and philosophical study*. London, UK: Rowan and Littlefield.

Fuller, M. (2003). *Behind the blip: Essays on the culture of software*. New York, NY: Autonomedia.

Fuller, M. (2008). *Software studies: A lexicon*. Cambridge, MA: MIT Press.

Galloway, A. R. (2004). *Protocol: How control exists after decentralisation*. Cambridge, MA: MIT Press.

Goodfellow, R. (2011). Literacy, literacies, and the digital in higher education. *Teaching in Higher Education, 16*(1), 131–144.

Goodfellow, R., & Lea, M. R. (2014). *Literacy in the Digital University: Critical perspectives on learning, scholarship and technology*. New York: Routledge.

Hawk, B., Rieder, D. M., & Oviedo, O. (2008). *Small tech: The culture of digital tools*. Minneapolis, MN: University of Minnesota Press.

Heim, M. (1987). *Electric language: A philosophical study of word processing* (2nd ed.). New Haven, CT: Yale University Press.

Hight, C. (2015). Software studies and the new audiencehood of the digital ecology. In F. Zeller, C. Ponte, & B. O'Neill (Eds.), *Revitalising audience research: Innovations in European audience research* (Vol. 5, pp. 62–79). New York, NY: Routledge.

JISC. (2014). *Developing digital literacies*. Retrieved from https://www.jisc.ac.uk/full-guide/developing-digital-literacies.

Johnson, S. (1997). *Interface culture: How new technology transforms the way we create and communicate*. New York, NY: HarperCollins Publishers.

Jones, C., Ramanau, R., Cross, S., & Healing, G. (2010). Net generation or digital natives: Is there a distinct new generation entering university? *Computers & Education, 54*(3), 722–732.

Kennedy, G., Judd, T. S., Churchward, A., Gray, K., & Krause, K.-L. (2008). First year students' experiences with technology: Are they really digital natives? *Australian Journal of Educational Technology, 24*(1), 108–122.

Khoo, E., & Cowie, B. (in press). Trial-and-error, Googling and talk: Engineering students taking initiative out of class. In D. Corrigan, C. Bunting, A. Jones, & R. Gunstone (Eds.), *Navigating the changing landscape of formal and informal science learning opportunities*. Berlin: Springer.

Khoo, E., Hight, C., Cowie, B., Torrens. R., & Ferrarelli, L. (2014). Software literacy and student learning in the tertiary environment: PowerPoint and beyond. *Journal of Open, Flexible and Distance Learning, 18*(1), 30–45.

Khoo, E., Johnson, E. M., Torrens, R., & Fulton, J. (2011). It only took 2 clicks and he'd lost me: Dimensions of inclusion and exclusion in ICT supported tertiary engineering education. In Y. M. Al-Abdeli & E. Lindsay (Eds.), *22nd Annual Conference for the Australasian Association for Engineering Education* (pp. 166–171). Fremantle, Australia: Engineers Australia.

Kirschenbaum, M. G. (2016). *Track changes: A literary history of word processing*. Harvard, MA: University Press.

Kitchin, R., & Dodge, M. (2011). *Code/Space: Software and everyday life*. Cambridge, MA: MIT Press.

Lankshear, C., Knobel, M., & Peters, M. A. (2016). *New literacies and digital epistemologies*. Retrieved from https://www.peterlang.com/view/serial/NEWLIT?v=toc.

Laurel, B. (1992). *Computers as theatre*. Reading, MA: Addison-Wesley Pub. Co.

Livingstone, S. (2004). Media literacy and the challenge of new information and communication technologies. *Communication Review, 1*(7), 3–14.

Livingstone, S. (2008). Engaging with media – a matter of literacy? *Communication, culture & critique, 1*(1), 51–62. doi: 10.1111/j.1753-9137.2007.00006.x

Manovich, L. (2001). *The language of new media*. Cambridge, MA. MIT Press.

Manovich, L. (2008). *Software takes command* (online draft). Retrieved from http://softwarestudies.com/softbook/manovich_softbook_11_20_2008.pdf.

Manovich, L. (2011). Inside photoshop. *Computational Culture: A Journal of Software Studies*, Issue One, available from http://computationalculture.net/article/inside-photoshop.

Manovich, L. (2012). *How to follow software users*. Available from http://manovich.net/content/ 04-projects/075-how-to-follow-software-users/72_article_2012.pdf.

Manovich, L. (2013). Software takes command. In *International texts in critical media aesthetics* (Vol. 5). NY: Bloomsbury Press.

Martin, A. (2008). Digital literacy and the "Digital Society". In Colin Lankshear & M. Knobel (Eds.), *Digital literacies: Concepts, policies & practices* (pp. 151–176). New York: Peter Lang.

Martin, A., & Grudziecki, J. (2006). DigEuLit: Concepts and tools for digital literacy development. *Innovations in Teaching & Learning in Information & Computer Science, 5*(4), 249–267.

McGrenere, J., & Ho, W. (2000). Affordances: Clarifying and evolving a concept. In *Proceedings of graphics interface 2000* (pp. 179–186), May 15–17, 2000, Montreal, Canada.

Oblinger, D. (2003). Boomers, gen-Xers, and millennials: Understanding the "new students". *EDU-CAUSE Review, 38*(4), 36–45.

Peeters, J., Backer, F. D., Buffel, T., Kindekens, A., Struyven, K., Zhu, C., & Lombaerts, K. (2014). Adult learners' informal learning experiences in formal education setting. *Journal of Adult Development, 21*(3), 181–192. doi:10.1007/s10804-014-9190-1.

Sefton-Green, J., Nixon, H., & Erstad, O. (2009). Reviewing approaches and perspectives on 'Digital Literacy'. *Pedagogies, 4*(2), 107–125. doi:10.1080/15544800902741556.

Selwyn, N., & Facer, K. (2007). *Beyond the digital divide: Rethinking digital inclusion for the 21st century*. Futurelab. Retrieved from http://citeseerx.ist.psu.edu/viewdoc/download?doi=10. 1.1.101.3384&rep=rep1&type=pdf.

Tapscott, D. (2009). *Grown up digital*. New York, NY: McGraw-Hill.

Tsatsou, P. (2017). Literacy and training in digital research: Researchers' views in five social science and humanities disciplines. *New Media & Society*. doi:10.1177/1461444816688274.

United Nations Educational, Scientific, and Cultural Organization (UNESCO). (2013). *Global media and information literacy (MIL) assessment framework: Country readiness and competencies*. Retrieved from http://unesdoc.unesco.org/images/0022/002246/224655e.pdf.

Valtonen, T., Dillon, P., Hacklin, S., & Väisänen, P. (2010). Net generation at social software: Challenging assumptions, clarifying relationships and raising implications for learning. *International Journal of Educational Research, 49*(6), 210–219. doi:10.1016/j.ijer.2011.03.001.

van Dijck, J. (2013). *The culture of connectivity: A critical history of social media*. Oxford, UK: Oxford University Press.

Vee, A. (2017). *Coding literacy: How computer programming is changing writing*. Cambridge, MA: MIT Press.

# Chapter 2
# A Genealogy of Software Applications

**Abstract** This chapter outlines a broad genealogy of two areas within software culture: Digital Non-Linear Editing (DNLE) and Computer-Aided Design (CAD) software. Emerging from distinct institutional environments, their respective historical developments and the implications these have generated within their professional domains provide a broader context for the software at the centre of this educational research project (see Chaps. 3 and 4). Each of these histories demonstrate how decisive the institutional and industrial contexts of their creation were in inscribing the affordances, interfaces and conceptual frameworks coded into these software.

## 2.1 Introduction

Every software application comes with its own history; it emerges from a particular context. The institutional context and corporate history for its emergence imparts a legacy, embedded within the code itself which evolves as the software becomes adopted by the user base and increasingly entrenched within professional and other practices. These contexts and histories are fundamental to the overall conceptual framework underlying the software. They also inform and shape its set of affordances (the set of actions possible within the software) and how these are organised in the design and configuration of such things as an application or platform's interface. As outlined in Chap. 1, these aspects of the software serve to enable and constrain the possible sets of practices to which such software can be applied by users. In a broader sense, software and its users evolve together, and a history of their development provides a necessary wider frame from the research 'snapshot' generated through our specific research project. The software explored in this chapter offer useful illustrations of how many applications developed in close partnership with professional practitioners, evolving into the software which later came to be used more widely, including becoming embedded into educational and training environments (the focus of our particular study).

The two disciplinary contexts we explore in our research are media studies and engineering (see Chaps. 3 and 4). This chapter begins with a (necessarily brief) narrative of the field of Digital Non-Linear Editing (DNLE) software, which is part

E. Khoo et al., *Software Literacy*, SpringerBriefs in Education,
https://doi.org/10.1007/978-981-10-7059-4_2

of a core set of media editors which have arguably transformed media production practices over the last 20 years. The second half of the chapter provides an overview of the development of Computer-Aided Design (CAD) software, which is central to a range of material practices, including those within the engineering discipline. The aim is not to provide an exhaustive history or genealogy of the specific software discussed in later chapters, but to suggest something of the trajectory of their development and outline the implications of their acceptance by practitioners within their respective creative fields.

## 2.2  The Development of Digital Non-linear Editing Systems (DNLE)

The long development of Digital Non-Linear Editing (DNLE) systems arguably represents a transformation as significant to moving images as word processing was for writing. However, this development is comparatively poorly researched as with much of *cultural software*, to use Manovich's term (see Chap. 1). In a sense, we are talking here about a *cut, copy and paste* approach to the construction of moving images. Earlier (analog) editing practices entailed a *destructive assembly* process, where a film strip was literally cut into pieces and reassembled until there was agreement on the final edit. While film could theoretically be endlessly recombined, in practice the materiality of film strips meant this gradually became more and more difficult. Digital systems, in contrast, allow for the random access retrieval of digital material in order to build a sequence that exists virtually, and with editing outcomes that are usually recorded (and outputted) as an edit decision list (EDL) (Murch, 2001). The key, and quite profound, advancement offered by DNLE systems, was that they allowed the creation of as many versions of film sequences as a user wanted as all that was being manipulated were digital files. The penultimate output from a DNLE system (the final EDL) was used as the blueprint to *cut* or *print* the film itself. The integration of this software into professional film making practices had enormous implications for production workflows, and arguably ultimately changed the ways in which audio-visual production came to be *imagined* by practitioners (as is discussed below). Despite the significance of this translation of the established practices of editing into software code, it is important to emphasise that film and television producers were by no means early adopters of incorporating computer-based tools into their production practices. The transition to a fully digitised production workflow faced many obstacles; some of these were technical as ambitions were delayed by the limitations of available technologies while others were more cultural and institutional. Innovations in this area happened first with small groups of early adopter professional practitioners, before slowly becoming more widely available as the cost of specialised editing systems decreased and editors became more accustomed to using these tools. The Moviola editor, a ubiquitous flatbed system for viewing/cutting film strips which had served as the standard analogue editing system since the introduction of sound in 1927, survived late into the 20th century, evidence of the decades it took for DNLE to achieve dominance within the industry.

The technical barriers to a completely digital process were not insignificant and generated caution from some of those who might have been early adopters. This reticence derived partly from the division between different physical formats, which made it difficult for developers of new versions of editing systems to recreate all editorial workflows. For example, there are key material differences between film and video editing (used respectively for film and television production) which meant that initially it was not possible for developers of editing systems to cater to both. The technical challenges of digitising editing practices also derived from the initial limitations of computer technology itself, such as storage and processing power. The distinction between offline and online editing,[1] for example, has gradually disappeared as available computers have become powerful enough to handle editing at full resolution. Some technical issues were too deeply embedded within existing production technologies to immediately solve, such as the distinction between the standards for European film and video (25 frames per second) versus US film and video (24 fps + 30 fps). These added costly conversions to and from film or video media files as part of production workflows—these challenges remain as legacies within file formats and compression codecs[2] today. Other challenges only emerged after earlier problems had been solved, and are part of a much broader and familiar trajectory of emerging technologies such as the transition to high definition (HD), 4K video and so on, which are beyond the scope of this overview.

It is important to emphasise that editing practices—that is, the ways in which editors conceptualised and organised their workflows—developed very much in tandem with each other in incremental ways that entailed a slow transformation of the nature of editing itself (Thompson, 1994). Rubin's narrative of this transformation highlights the key period of 1989–1993 as the real emergence and dominance of fully digitised NLE (Rubin, 2000), but there are complex, overlapping developments in this history with most periods characterised by long time lags as established technologies survived even when there was rapid development of digital technologies. Even today there are some film directors who insist on creating motion pictures on film despite the process of distribution and exhibition having been largely digitised. There are a number of potentially significant (but largely under-theorised and under-researched) milestones in the slow reconceptualization of audio-visual editing as a coded practice (see Dancyger, 2011; Ohanian 1998; Rubin, 2000; Thompson, 1994). The year 1995 is frequently cited as a key year in which digital editing became more widespread within the more elite editing practices (and budgets) of Hollywood production. This is marked especially with the dominance of the Avid Film Composer system which epitomised developments that had been made in solving key technical challenges,

---

[1]Offline editing involved transferring film to video, to make it easier for editing systems to deal with the digitised footage.

[2]A compression codec (short for coder-decoder) encodes a media file for storage and distribution, and decodes it for playback or editing.

including the problem of video/film footage transfer. Together with Apple's Film Cut Pro and Adobe Premiere, these constituted the big three of professional level editing systems. Most crucially for this history, the unbundling of software from hardware systems fostered a more competitive environment which resulted in the rapid widespread adoption of any innovations pioneered by any one vendor. With the increasing competition between a small number of key players, innovation became secondary to standardisation, and this is most evident in the emergence of the now familiar template for editing interfaces.

## 2.2.1   The Interface

The graphic user interface (GUI) for digital non-linear editing—the interface which editors engage with on their computer screens—is the what confronts users when opening any audio-visual editing application, and the elements of its basic design and key features are replicated across both professional-level applications and those designed for more novice users. In its layout and terminology, this interface retains something of the legacy of the physical operations of film flatbeds (such as the Moviola system). The various elements of this interface gradually came together in successive iterations of DNLE applications produced by various vendors over a numbers of years. Ohanian provides a useful summary of the key elements of the contemporary DNLE system (Ohanian, 1998, pp. 52–56):

- The Clip: the granular component of all editing, derived from *the shot* in film editing, which tends to represent a single continuous set of footage, and which is represented by an icon, text, and frame in the interface.
- The Transition: derived more videotape editing, where there was more of a need to *fill in* the space left when a shot was trimmed, but now appearing as a variety of options for editors to apply across cuts between clips.
- The Sequence: a sequential series of (trimmed) clips, stills and other material (such as various kinds of audio), which can in turn be combined into larger sequences and so on. These sequences, a key building block for editing, might be generated by different editing teams on large productions, and combined later.
- The Timeline: the centre of the interface, where multilayered sequences (combining layers for different video and audio material) can be combined to create sequences which play out over time.

These essential elements of this interface have been replicated across editing software and have become deeply integrated with other applications to form the basis of a profoundly transformational approach to the construction of moving-image media content. This is how editing is now performed and imagined by editors (and other personnel in the production line).

## 2.2.2 The Implications of DNLE

The provision of fully-digital workflows, fostered by software from major players such as Avid, Apple, and Adobe, prompted an immediate reorganisation of production workflows. For many large productions, instead of a single editing team providing a single bottleneck for all raw footage, there might be a number of machines operating simultaneously, working on different footage, to be combined later.

The implications of DNLE have been more profound than this; however, the extent to which *editing* has been redefined through digital workflows is heavily debated within industry circles. Walter Murch is one of the few practitioners who has reflected on the transition from analogue to digital editing practices. His seminal text *In The Blink of An Eye* (2001) reveals his enthusiasm as an early adopter of digitised editing across the film industry. His perspective is an informed one, and a good illustration of the close and dialectical relationship between professional practitioners and the evolution of the systems they adopted.

Murch also appears in an interesting case study on the transition to digital editing practices: Koppelman's (2005) account of Murch editing *Cold Mountain* (Minghella, 2003) on Apple's Final Cut Pro (FCP) system. This was a production using film on location, then the footage was migrated to digital video and came out the other end as film again for distribution to theatres (with film distribution yet to convert to digital projection as the norm). Significantly, FCP was considered, even by Apple itself, as a *prosumer* application, midway between professional and consumer software. FCP was widely used to edit documentaries, but not considered robust enough for feature film production. Murch's adoption of the system on a big budget feature film represented both an early example of the convergence of consumer-level digital video tools and Hollywood film industry, and a fascinating account of a software developer being pushed to re-imagine an application developed exclusively for digital video, and which in fact required third-party tools to operate (Koppelman, 2005).

These kinds of accounts reveal how practice evolved in tandem with software development. Murch's reflections on the nature of DNLE in his own writing, and as relayed by Koppelman in his case study, provides a useful summary and reflection of broader opinion within editing practitioners as a community. Overall, although this is expressed in different ways, there is a recognition that the shift from the materiality of film, from the *destructive* approach of editing film, to a fully digitised system has meant the adoption of a new conceptual approach. This has played out in different ways for different kinds of editing practice, but there are some broad observations articulated by most editors.

DNLE allows for the possibility of increased speed in editing, which itself means a less considered and methodical exploration of the potential ways to combine and recombine clips into sequences. This generally means a more efficient editing process (and hence less costly, a key driver in the adoption of these systems). These efficiencies are somewhat mitigated within digital workflows by the vastly increased volume of footage which is able to be captured using digitised cameras with large storage capacities. As DNLE allows a system to develop and create multiple versions of an edit, these workflows have also opened up the editing room to the more direct intervention of directors and other personnel. DLNE also allows for a more integrated approach to how image and sound might work together, rather than the older production process of adding sound (Murch, 2001). As Dancyger notes, individually these are all small changes in workflow and the ways in which editing is imagined, but collectively they represent a significant change in practice (Dancyger, 2011).

Murch himself highlights the changes to the nature of analog film editing as a physical, embodied practice. His own practice involved standing at a Moviola flatbed deck, using his whole body to work the viewing and the cutting of film strips in a way that became intuitive (Murch, 2001). He notes that editing involves the logistical wrangling of footage, analysis of the structure of sequences together into a rough edit, and the actual performance of the editing itself, and all three areas are transformed within DNLE (Koppelman, 2005). He also developed specific elements to his workflow that are only possible with the materiality of film; for example, he would physically rewind the editing tape back to the beginning through the viewfinder, meaning he would watch sequences backwards to get a completely new perspective on its structure—something that isn't possible with the *scrubbing* feature of digital video players (which allows users to jump ahead multiple frames, to skip through sequences at high speed).

The overall speed and ease of this cut, copy and paste approach to editing attracted complaints from some practitioners and commentators that this has degraded the quality of the considered reflection that needs to be at the heart of distinctive and innovative editing solutions for each project (Murch, 2001). These accounts point to the emergence of more formulaic and standardised approaches to editing across different kinds of media content as a key implication of digitised workflows. Ellis (2012) argues that accelerating the process of editing has had implications within broader patterns of accelerated cutting in media content, something he characterises as a loss of craft and individual editing styles (Ellis, 2012) and a greater density in cutting styles, such as quicker cutting between multiple perspectives and angles within the same scene.

Perhaps the most profound transformation associated with DNLE, however, is not provided by the affordances of the systems themselves, but facilitated by the ease with which material can be imported and exported to other forms of software. Now it is often the case that different pieces of software will handle specific kinds of image and sound construction and editing, which are then combined as layers within a more generic media editing application. This broader context of exchange of digitised material means that *coded* filmmaking processes have taken on very different qualities to previous eras, and this is manifest in the types of changes exhibited within media content more generally.

Manovich's analysis of the Adobe After Effects (AE) application is a useful addition to debates in this area (AE is part of the package which is taught within universities as industry-standard, including within the media discipline researched within our project, see Chap. 3). Manovich writes as a practitioner, noting the changes to his own practice, and highlights the period 1993–98 where a change in the aesthetics of particular kinds of media content became noticeable. He uses the term Velvet

Revolution (as in the slow drift of revolution in Czechoslovakia in 1989) to describe this gradual transformation, led by AE and a small number of similar programmes, which have fostered a new *hybrid visual language* of motion graphics (Manovich, 2006).

> What is the logic of this new hybrid visual language? *This logic is one of remixability: not only of the content of different media or simply their aesthetics, but their fundamental techniques, working methods, and assumptions.* United within the common software environment, cinematography, animation, computer animation, special effects, graphic design, and typography have come to form a new 'metamedium'. A work produced in this new metamedium can use all the techniques which were previously unique to these different media, or any subset of these techniques (Manovich, 2006, p. 10, emphasis in original).

Instead of creating films where an animation sequence was followed by a live action sequence and so on, these various kinds of media content (generated by quite different workflows and raw materials) could all operate as layers within a single overarching timeline, and ultimately begin to interact at a more fundamental level. It is only over time that he belatedly recognised, even as a practitioner, the implications for his own imaginative possibilities for creating media content, as motion graphics began to become the standard for short-form audio-visual content, such as television commercials, and the opening credit sequences to television programmes.

DNLE itself is now typically packaged within a larger ecosystem of production tools, all specialised media editors which are increasingly imagined to operate together to provide a wide spectrum of possibilities for media producers to play within. The emergence of these software-based tools suggests a redefinition of audio-visual practice itself, their collective impact and the emergence of distinctive new conceptual frameworks. Underlying these changes in organising screen content are greater uncertainties concerning the organisation of creative labour itself:

> Specifically, digitization has facilitated a collapse and confusion of production workflow and upended traditional labor [sic] hierarchies. Workflow refers to the route that screen content travels through a production organization and its technologies as it moves from the beginning (origination, imaging, recording) to the end (post-production, mastering, duplication, exhibition) of the production/distribution process. [...] In fact, the once linear sequence through which filmed material went before being printed and broadcast has fallen apart. Because of these recent shifts to digital, visualization and effects functions once reserved for post-production now dominate production, and skills once limited to production now percolate through post-production (Caldwell, 2011, p. 293).

A host of software-enabled specialisations, such as colour grading, motion capture, the generation of CGI, and motion graphic techniques, allow for a wider palette of techniques available to media producers. Admittedly, this is a narrative which does not encompass all of audio-visual production; at the opposite end of the filmmaking spectrum are mobile, amateur and networked practices which have reconfigured DNLE in quite different directions (Hight, 2014a, b). Overall, however, the students participating in our research encounter sophisticated, professional-level editing systems with specific conceptualisations embedded in their hierarchy of affordances and in their interfaces. Next, we turn our attention to another ubiquitous software, in this case used to facilitate the design of engineering, architectural and other physical arte-facts in three-dimensional (3D) format. Computer-Aided Design (CAD) is another part of software culture which has had wide-ranging implications for a reimagining of creative practices across a number of related industries.

## 2.3   The Development of Computer-Aided Design (CAD)

As with the discussion on Digital Non-Linear Editing (DNLE), what follows is necessarily truncated and cursory, as we do not have the space here to delve into the wealth of literature which attempts to analyse and summarise the implications of Computer-Aided Design (CAD) practices. In our own small project we engaged specifically with an engineering discipline, but it is important to note CAD is an aspect of software culture with wide applications within design, architecture and related practices, where it has become a given set of tools with wide-ranging implications for the nature of professional practice.

CAD involves the use of software in the creation, modification, analysis or opti-misation of material design (if we define this broadly, to include a range of practices from the design of nuts and bolts, through to more complex forms of mechanical engi-neering encompassing everything from automotive to bridge design, and ultimately to forms of built environments or architecture). Some of the transformation of mate-rial practices associated with this kind of software has been extensively debated, particularly within architectural literature. This befits a field which sees itself as aspects of design practice which transcend the merely functional. In these circles digitised workflows consequently attracted intense debate over the social, cultural and political implications of its outcomes.

CAD arose from a very different institutional environment to DNLE (with the Massachusetts Institute of Technology playing an outsize role), but there are some parallels and interesting points of comparison in terms of the significance of a number of new conceptual frameworks which emerging software eventually come to embody and foster. We are concentrating on architecture and engineering in this account, but there are obvious areas now where 3D modelling and media editing software operate together within particular kinds of creative practices (the most recent and celebrated include augmented and virtual reality, but there are deep roots here into forms of computer graphics and game design). As with all software it is increasingly obvious that applications and platforms formed within one sector of human endeavour quickly start to become part of the broader incestuous and prolific combinatorial evolution of software culture (as broadly outlined in Chap. 1).

The development of CAD forms one part of a broader history of engineering and architectural design practice itself, and is associated with a number of transfor-mational milestones in these practices. Some of the earliest *technical* drawings for machines or devices date back to the 14th or 15th century, among the most famous are those produced by Leonardo da Vinci. However, if we were to consider these draw-ings in a modern context they would be described as *sketches* as they lack dimensions or scales and often have exhaustive text descriptions to help the viewer understand the intent (Weisberg, 2008, p. 2–1). These early drawings served two purposes: a reference for skilled craftsmen to construct the device depicted and also as a portfolio to present one's work to a wealthy patron (Lefèvre, 2004). Crucially, at this point in history there was a clear separation in practice between those who offered designs of material objects and those who actually built such things based on those designs. In marked contrast to contemporary practice, this was not a collaborative relationship nor a space where early *architects* were acknowledged as the drivers of projects.

Leon Battista Alberti is invariably credited with inventing modern architecture, in the sense that he exploited the new technologies to insist that the designer was the author of a building and no longer beholden to the craftspeople who actually created a building. Before Alberti, architects had to contend with builders who interpreted their designs according to their own practices and the demands of their local contexts. So the creation of a building was an inherently collective and decentralised process, relying on oral, material and technical traditions outside of the control of the architect (Llach, 2015). The Albertian paradigm is a key reference point for understanding the emergence of CAD. One of the broader ironies of this history is that this software at first seemed to fulfil the promise of the Albertian approach, but has in more recent years gradually undermined it.

Using a new notational system, and exploiting the possibilities for the new technology of print to provide an exact replica of a design, Alberti could insist that the architect was indeed the *author* of a building, not just a starting point for a design which was re-shaped on location by other craftspeople. So in Alberti's terms, "the *design of the building is the original, and the building is its copy*" (Carpo, 2011, p. 26, emphasis in original). Following Alberti, the notational system of architecture helped to establish a distinct identity for architecture, which in turn eventually helped to set the conceptual stage for the arrival of computers as tools to serve these masters (Llach, 2015).

The specific origins of CAD are typically seen as located within the Massachusetts Institute of Technology (MIT), which produced the first CAD software, Sketchpad, as part of Ivan Sutherland's doctoral research in 1959,[3] building on a variety of earlier work by researchers inside and outside the institution (Cohn, 2010; Llach, 2015). As with early DNLE development, there were very few practitioners who had the resources to commit to investigating the use of the early prototypical and expensive systems. Consequently, the early development of CAD, in an engineering context, was primarily driven by large aerospace and automotive companies. These were companies which were able to afford the expensive computer equipment required and were already engaged in such complex design processes that they were attracted to the possibilities of the reduction of drawing errors, increased reusability of drawings and greater efficiencies promised by CAD. It is important to recognise that the adoption of these systems was driven by a search for greater efficiencies in productivity rather than a design tool. Instead they offered systems to find drawings more quickly,

---

[3] Not coincidentally, Sutherland later emerged as a key figure within the history of computer graphics.

simplify modifications of drawings and allow the automation of some parts of drawing practices (CADAZZ, 2004).

As Llach notes, in contrast to the popular conception that Computer-Aided Manufacturing (CAM) is an offspring of Computer-Aided Design (CAD), the opposite is true. Like filmmaking, engineering and architecture were comparatively late adopters of embedding computers into everyday creative practice. CAD developed from experiments to automate manufacturing, and it was only later that the transformational potential for design itself came to be realised (Llach, 2015).

The ethos and vocabulary of manufacturing gave origin to the first CAD systems (Llach, 2015, p. 37), but this was also, unusually for software culture, a highly theorised process. The development of CAD at MIT was complex, and significantly involved a great deal of debate about the nature and desirability of the human-machine hybrid practice which might result. MIT not only developed CAD as a tool, but generated a series of accompanying theoretical reflections that helped to shape assumptions on how it might operate within industry. These debates centred on the use of creatively using computers, the need to divide labour between humans and machines and the implications of re-imagining material design as a kind of data processing (Llach, 2015).

These debates are quite distinctive from those associated with the development of DNLE, as they drew upon a broader caution about the nature and role of computers within material design practice. The term *Computer-Aided Design* itself reflects the demand that computers support human creativity, rather than any sense that there should be a collaboration between human and machine (Llach, 2015). CAD consequently tended largely toward generating efficiencies through the augmentation of pre-existing practices in the 1970s and 1980s (Llach, 2015). For example, engineering within aerospace leader Boeing had an *all CATIA, no paper*[4] design strategy. This led to a substantial reduction in time to market by safely eliminating the need for physical mock-ups (often required to verify paper designs). The typical impetus for the adoption of CAD was still the quest for workflow efficiencies. In late 2000 automotive manufacturer Ford showed that 3D CAD, with internet enabled product data management (PDM), could cut the concept to shelf time to approximately one third of that required by the more common, non-internet enabled techniques. The primary advantages of the network enabled method were that they allowed viewing and collaboration by geographically dispersed teams on a single digital master, almost eliminating the misfit and mismatch problems often associated with globally dispersed manufacturers and parts suppliers.

While MIT was crucial to the broader development of CAD, and succeeded in actively shaping the popular imagination with design fields and wider (Llach, 2015), ultimately how CAD developed diverged from this original role. The longevity of its introduction into everyday practice perhaps aided in this adaptation, as CAD gradually drifted further from how it was conceived by its creators, as it became diffused through architectural and engineering practice. As the software itself became more sophisticated with enhancing computer technologies, there was a gradual shift of

---

[4]CATIA is a CAD platform.

focus from simply automating the practice of drafting into something more trans-formative: the emergence of a platform facilitating a comprehensive building (and design) simulation (Llach, 2015). These are all developments which at first glance appear to provide a narrative of inevitable transformation, a confirmation of the claims of technological determinism. Initial CAD programs effectively just trans-lated the blueprinting process onto a digital platform, and it was only as the software increased capability to allow for the techniques of 3D modelling that its broader creative capacities became prevalent.

Modern 3D CAD programs include a variety of sophisticated analysis tools that allow various simulations to be run on the 3D item/structure. This has given rise to the term *virtual product development*, where products are developed and prototyped in an entirely digital medium (CADAZZ, 2004). Today, CAD is used extensively in most activities in the design cycle, everything from recording product data, to allowing for remote collaboration between design teams (Bilalis, 2000). The open co-creation possibilities of CAD software emerged gradually, but also in a highly theorised way, a reflection of the significance of the university environment as a breeding ground for its conception and early prototypes.

Interestingly, Llach's critical perspective draws explicitly from the Software Stud-ies paradigm, arguing that software needs to be examined "as part of the infrastruc-tures that condition the design and production of built environments" (Llach, 2015, p. 23). For commentators such as Llach, what was at stake is the nature of the creative endeavour itself. Before the widespread use of CAD in the education of engineers, there was much greater emphasis on drawing and sketching (Buchal, 2002). Hare (2005) says that sketching is inherently creative, the practicing and sketching fre-quently leads to more creative thinking; in fact, analog tools, such as pen and paper, are still viewed as more haptic and intuitive. From this perspective, CAD can guide an engineer through technical issues, such as dimensions and scaling, but it does not have the same ability to create quick visualisations like sketching does. Moving to a CAD workflow, then, might mean losing key elements of design practice.

Llach cautions against making generalisations in this area however, as the use of the CAD tool has varied greatly and the use of these systems are deeply informed by practitioners' own position within debates over the role which computer-based practice should play. He cites the example of Frank Gehry, who continued to construct actual models, which were then scanned and inserted into computer form. The process is complex here, as the potential of the software also clearly informed the imagination of architects, allowing them conceptual space to potentially re-imagine the nature of their own practice. The role of the software is still, in everyday disciplines globally, negotiated and framed by broader agendas and localised practices (Llach, 2015).

### 2.3.1 Implications of CAD

The overall paradigmatic changes associated with CAD workflows have been neither universal nor linear. Initially this software represented a confusion of the Albertian perspective, and the emergence of a vision of architectural design as data processing (Llach, 2015, p. 66), in the process "revealing software as a territory where the meaning of design itself is negotiated" (Llach, 2015, p. 87). Rather than a slave for the Albertian paradigm, the computer has sparked a profound refashioning of the nature of material design practices, such as engineering, with debates now centred on the nature of the human-machine assemblage that has emerged, and which way development should now progress.

Just as DLNE is now part of a broader production ecology that challenges understandings of what media are (Manovich as cited in Chap. 1), Llach argues that software is a site for competing theorisations about design, and consequently, "the technology project of CAD appears as a disciplining project, not an emancipatory tool, but rather a governing one" (Llach, 2015, p. 102). The broader implications are complex, and there is (again) notably more detailed and extensive theorising about these aspects within discourses surrounding CAD than for DNLE.

Robertson and Radcliffe (2009) argue that "there is growing evidence that the ubiquitous CAD tools that design engineers use in their everyday work are influencing their ability to solve engineering problems creatively, in both positive and negative ways" (p. 136). Positive factors include the ability to visualise and *play* with designs, less time spent on detail (potentially allowing more time on being creative), and enhanced communication facilitating *group creativity*. Negative impacts tend to be vaguer, though Robertson and Radcliffe have identified four general categories:

- Enhanced visualisation and communication: there are obvious positive aspects to this category. Negative impacts included having clusters of people crowding around a monitor hampering brainstorming; and the tendency when a detailed CAD model was displayed for it to convey an illusion of completeness and discourage further creativity.
- Circumscribed thinking: this could either be where the functionality of CAD limited solutions (either to what was possible to do in CAD, or perhaps worse, what was easiest to do in CAD); or at the other end of the scale very proficient CAD users using the functionality of the tool to develop unnecessarily complex designs because CAD allowed it rather than because these were the best design solutions.
- Bounded ideation: the notion that using CAD for large portions of a day was not necessarily conducive to creativity (the mundane nature of drafting along with technical problems and software bugs being a distraction from the process of designing).
- Premature fixation: as CAD models became more complex (usually as the design process proceeds) there was greater disincentive to make changes (presumably due to the amount of work that would be required to make these).

As always, debates centred on whether such new human-machine assemblages truly enhance innovative and effective design practice. Some commentators insist on a profound paradigmatic change prompted by CAD, with hints of the technological determinism underlying some writing on software culture more broadly. Carpo writes that the "Albertian paradigm is now being reversed by the digital turn" (Carpo, 2011, p. 27).

The idea that the new digital design tools could also serve to make something else – something that would not otherwise have been possible – may have occurred when architects began to realize that computer-aided design could eliminate many geometrical and notational imitations that were deeply ingrained in the history of architectural design. Almost overnight, a whole new universe of forms opened up to digital designers. Objects that, prior to the introduction of digital technologies, would have been exceedingly difficult to represent geometrically, and could have been produced only by hand, could now be easily designed and machinemade using computers (Carpo, 2011, p. 36).

There is parallelism here with the developments of fully realised CGI-animated film worlds fostered by media production, but prompted also by the influence of postmodern theorists such as Gilles Deleuze. He offered a new language of *folds* in architectural design (helping to prompt the development of algorithmic affordances in CAD platforms which could realise these in virtual form). The fold, "a unifying figure in which different segments and planes are joined and merge in continuous lines and volumes, is both the emblem and the object of Deleuze's discourse" (Carpo, 2011, p. 86).

And, crucially, the rigour of the Albertian paradigm is much more compromised now within this environment. Instead of a firm commitment to the authorship of the architect, who produced a design and anticipated that it would be exactly replicated in the building itself, and the early CAD phase where the software was used to implement broader assumptions of standardisation and automation, today CAD allows for a more fluid and ever changing re-imagination of the nature of design itself. An architect's original plan could once again (as in pre-Alberti times) be endlessly reinterpreted using individual explorations at different points in the design process: "In a digital production process, standardization is no longer a money-saver. Likewise, customization is no longer a money-waster" (Carpo, 2011, p. 41).

As CAD became more embedded within material design disciplines, such as engineering, it allowed three-dimensions to become part of the authoring process. As with DNLE the broader ecology of software development has

made it possible to envisage a continuous design and production process where one or more designers may intervene, seamlessly, on a variety of two-dimensional visualizations and three-dimensional representations (or printouts) of the same object, and where all interventions or revisions can be incorporated into the same master file of the project. This way of operating evokes somehow an ideal state of original, autographical, artisanal hand-making, except that in a digitized production chain the primary object of design is now an informational model (Carpo, 2011, p. 33).

A key point was that design representations now became "forms of *building*" structured information, engineered rather than drawn (Llach, 2015, p. 67, emphasis in original). Architects were able to model new constructions in the software itself. A key shift here is toward the term "*modeling*, often used by architects to describe the production of three-dimensional descriptions in software, [which] evokes manual work in a way that other words, such as simulation, do not" (Llach, 2015, p. 100). The CAD process has evolved increasingly into an ever-more data-intensive set of practices, with recent developments of building information modelling (BIM) deeply embedded within automated practices allowing and requiring greater databases of content to be folded into the design process.

The CAD systems currently available for students within disciplines such as engineering (one of the foci of our research in the following chapters), then, are highly sophisticated, densely designed software-based platforms enabling a wide variety of material practices. They have evolved from their origins as tools to serve, support and help implement human creativity, emerging as human-software engines which are challenging for any practitioner to master, and typically offer a daunting software environment for novice users to encounter (as we shall see in Chap. 4).

## 2.4  Summary

This chapter has scoped the genealogy and development of two distinctive forms of software—DNLE and CAD—commonly taken up within the professional fields of media studies and engineering today. Obviously, it is not enough to provide such broad histories, as they reveal little but generalisations and theorising extemporised from exemplars and case studies at hand. What is required from this point is more detailed explorations of how and the extent to which these might play out within specific institutional contexts, whether these broader patterns and generalisations hold true across different practices, deployed by distinct practitioners, within institutional variations, and any number of other factors. We turn to this task in the next two chapters.

## References

Bilalis, N. (2000). *Computer aided design CAD*. Report produced for *INNOREGIO: Dissemination of innovation and knowledge management techniques*. Retrieved from http://www.adi.pt/docs/innoregio_cad-en.pdf.

Buchal, R. O. (2002). Sketching and computer-aided conceptual design. In *7th International Conference in Computer Supported Cooperative Work* (pp. 112–119). Design, Brazil: IEEE.

CADAZZ. (2004). CAD software—history of CAD CAM. Retrieved from http://www.cadazz.com/cad-software-history.htm.

Caldwell, J. T. (2011). Worker blowback: User-generated, worker-generated, and producer-generated content within collapsing production workflows. In J. Bennett & N. Strange (Eds.), *Television as digital media* (pp. 283–310). Durham, NC: Duke University Press.

Carpo, M. (2011). *The alphabet and the algorithm*. Cambridge, MA: MIT Press.

Cohn, D. (2010). *Evolution of computer-aided design, digital engineering* (December 2010). Retrieved from http://www.digitaleng.news/de/evolution-of-computer-aided-design/.

Dancyger, K. (2011). *The technique of film & video editing: History, theory, and practice* (5th ed.). Burlington, VT: Elsevier.

Ellis, J. (2012). *Documentary: Witness and self-revelation*. London, UK: Routledge.

Hare, R. (2005). *The act of sketching in learning and teaching the design of environments: A total skill for complex expression*. Retrieved from http://www.lboro.ac.uk/microsites/sota/tracey/journal/edu/hare.html.

Hight, C. (2014a). Shoot, edit, share: Cultural software and user-generated documentary practice. In K. Nash, C. Hight, & C. Summerhayes (Eds.), *New documentary ecologies: Emerging platforms, practices and discourses* (pp. 219–236). Houndmills, UK: Palgrave Macmillan.

Hight, C. (2014b). Automation within digital videography: From the Ken Burns effect to 'meaning-making' engines'. *Studies in Documentary Film, 8*(3), 235–250.

Koppelman, C. (2005). *Behind the seen: How Walter Murch edited Cold Mountain using Apple's Final Cut Pro and what this means for cinema.* Berkeley, CA: New Riders.

Lefèvre, W. (2004). *Picturing machines 1400–1700.* Cambridge, MA: MIT Press.

Llach, D. C. (2015). *Builders of the vision: Software and the imagination of design.* New York, NY: Routledge.

Manovich, L. (2006). After effects or the velvet revolution. *Millennium Film Journal, 45*(46), 5–19.

Minghella, A. (2003). *Cold Mountain.* Miramax. DVD.

Murch, W. (2001). *In the blink of an eye: A perspective on film editing* (2nd ed.). Beverly Hills, CA: Silman-James Press.

Ohanian, T. A. (1998). *Digital nonlinear editing: Editing film and video on the desktop.* Boston, MA: Focal Press.

Robertson, B. F., & Radcliffe, D. F. (2009). Impact of CAD tools on creative problem solving in engineering design. *Computer-Aided Design, 41*(3), 136–146.

Rubin, M. (2000). *Nonlinear: A field guide to digital video and film editing* (4th ed.). Gainesville, FL: Triad Publishing Company.

Thompson, C. (1994). *Non-linear editing: A survey.* New Technology and Training Series. London, UK: Skillset, The Industry Training Organisation for Broadcast, Film & Video and British Film Institute.

Weisberg, D. E. (2008). A brief overview of the history of CAD. In *The engineering design revolution: The people, companies and computer systems that changed forever the practice of engineering,* (pp. 2–1). Available at http://www.cadhistory.net/02%20Brief%20Overview.pdf.

# Chapter 3
# The Learning, Use and Critical Understanding of Software in Media Studies

**Abstract** This chapter (as with the next, Chap. 4) reports on the findings from a two-year funded empirical study (2013–2014) exploring how tertiary students in media studies and engineering develop the understandings and skills needed to use software as forms of software literacy. Two case studies were developed. The case studied experiences of media studies students' software literacy development is the focus of this chapter. Two cohorts of media studies undergraduate students were tracked, at different stages of study and using mixed methods, in their learning of discipline-specific software, Final Cut Pro, and the Adobe Creative Suite. The findings illustrate the ways student software literacy develop in a specific tertiary context. The findings will be revisited in Chap. 5 and discussed to include implications for the wider field of software teaching and learning.

## 3.1 Introduction

As highlighted in the first chapter, there is increasing expectation that media studies students will be able to engage competently and critically with the wide array of creative media software available in the industry. The rise of digital technologies have reshaped the nature of media studies as a discipline blurring the lines between those who are producers and consumers of digital content (Manovich, 2006). In other words, the boundaries between consumers and producers are being broken down by increased accessibility to technology and software. Yang (2014) argues that this is leading to a fusing of producer with consumer, and a new term, *prosumer*, may be more apt to describe future digital users. The *home user* or the *creative maverick* are thus increasingly seen as valid outcomes for media literate graduates. More specifically, engaging with creative media software, either as a teaching or learning tool, is seen as an essential and almost unavoidable part of modern learning, particularly when preparing students for adult life. Some go as far as to argue that creative media software (Adobe Creative Suite, Final Cut Pro, Paintshop, etc.) should be taught to students as early as high school, with many other researchers (e.g., Livingstone, Wijnen, Papaioannou, Costa, & del Mar Grandio, 2014) arguing that learning centred on creative media software should be viewed as an essential literacy,

© The Author(s) 2017

E. Khoo et al., *Software Literacy*, SpringerBriefs in Education,
https://doi.org/10.1007/978-981-10-7059-4_3

i.e., media literacy, additional to traditionally taught literacies, such as reading and writing.

However, a substantial research gap still exists concerning how students learn, engage with and view discipline-specific creative media software. Few studies have explored the impact of informal learning networks in relation to learning within media studies or, similarly, the effects these approaches are having on formal learning networks. Most studies have concentrated on notions and developments of digital information literacy, or information literacy skills. See, for example, Hegarty et al. (2010) who provided a progression for the development of digital literacy skills based on students' ability to access and evaluate electronic information in order to critically manipulate and use such information for their learning purposes in recognition of the broader social and cultural contexts within which the information is situated. The study, however, failed to examine the nature of student critique and decision making around which tools might best serve their purpose. Developing the ability to critique constitutes an essential characterisation of a 21st century learner (Gilbert, 2005).

This chapter reports on a study which aimed to explore, examine and theorise on how the notion of software literacy is understood, developed and applied in tertiary teaching and learning contexts, and the extent to which this understanding is useful when translated into new contexts of learning with and through software. We view this understanding as crucial and relevant to ensure all students and lecturers are better supported in teaching and learning processes that are mediated through and focused on software. Sociocultural theoretical perspectives are adopted as a basis of our study. The chapter begins by visiting the research design which case studies the experiences of undergraduate media studies students before detailing the case and expanding on the findings as evidence for the ways students engage with discipline-specific software.

## 3.2   Research Design and Context

This research draws from data collected in a two-year longitudinal research project funded by the New Zealand Ministry of Education: Copy, cut and paste (CCP): How does this shape what we know? (Khoo, Hight, Torrens, & Cowie, 2016), to report on the views of participating tertiary media studies and engineering students from a New Zealand university. The research questions guiding the investigation were:

1. To what extent, and how, does student software literacies develop and impact on the teaching and learning of discipline-specific software in formal tertiary teaching settings?
2. What software literacy do students consider they learnt as part of the case study tertiary course(s)?
3. How and in what ways do lecturers model attention to and use of different aspects of software affordance in a course which utilises discipline-specific software?

The research intention was to unpack if and how students develop and use discipline-specific software literacy, understand the influence of software on the way they make sense of disciplinary knowledge and whether their learning trajectories fit with the hypothesised tiers of software literacy (see Chap. 1 for a discussion of the three-tier software literacy framework).

With reference to digital technology adoption and use, Selwyn (2010) asserts that the importance of developing deep understandings of local contexts and ICT practices cannot be underestimated. A qualitative interpretive methodology was thus adopted to frame the study as it is consistent with the intention of uncovering the significance of events as experienced by research participants (Bell, 2004; Maykut & Morehouse, 1994) so that worthwhile improvements to learning can occur. It is congruent with a sociocultural framework that values the social and cultural contexts for how knowledge is co-constructed through interaction between individuals and tools (Wertsch, 1998).

A case study approach was adopted to allow the research team to develop an in-depth understanding of participants' lived experiences and transformations throughout the period of the study (Gall, Borg, & Gall, 1996). Two information rich cases were purposively selected (Patton, 2002) from two diverse disciplines (media studies and engineering) within a New Zealand university—the University of Waikato. Their selection was in part based on collaboration with lecturers who were keen to examine the notion of software literacy.

## 3.2.1 The Research Design

An overlapping longitudinal study design (also known as cross-sequential or patched-up design (Arzi, 1988) was adopted to track shifts in equivalent student cohorts' software literacy development. This design provides the unique opportunity and advantage for studying change over the long term when a full longitudinal study is not practical (Arzi, 1988). This design was a practical alternative to a longitudinal study of the same students over the full three or four years of their tertiary programme. It enabled the research team to map student learning and development across the entire three or four years of a degree programme in two years. In this way, the tiers in the hypothesised software literacy framework was tested across the year levels.

## 3.2.2 The Media Studies Case

The media studies programme generally attracts a large number of enrolments (100–150 students) from students with diverse backgrounds at the entry level. Students are required to engage with discipline-specific software through laboratory-based work and individual and group-based projects as well as provided with resources for informal learning. This case study involved collaborations between the research team and media study lecturers who were keen to examine the notion of software literacy through the teaching and learning of commonplace discipline-specific software—Final Cut Pro, After Effects, and Adobe Creative Suite (a range of media editing software). Hence within the case, a range of courses focused on the teaching of disciplinary software were investigated.

Within the media studies first year course, students are generally young school leavers with 10–15% of these being international students. In the first year of the media studies case, there were some students who have studied the software package in a previous course while others have had direct entry to the course. The extent media studies students can demonstrate advancing software literacy skills and how this plays out in terms of their interpretation of their discipline knowledge was of interest in this study. The research team therefore tracked one group of students from Year 1 into Year 2 and another group of students from Year 2 to Year 3 courses. The assumption is that, on the whole, students' software literacy develops as they gain more experience with a particular software package although a linear progression was not assumed.

### 3.2.3  Data Collection

Multiple data were collected to address the research questions through:

- lecturer individual interviews and tutor focus group interviews (up to four per course) to obtain lecturer/tutor perceptions and awareness of the affordances of their discipline-specific software and how this influenced the teaching and learning of the software,
- observations of lectures and laboratory (lab) sessions (up to two observations per course) to understand students' learning to use discipline-specific software,
- online student surveys to obtain general student evaluation on the teaching and learning of software at the end of each course. The survey (see Appendix 3.1) consisted of three sections asking students about their:

  - demographic profile,
  - software experience prior to enrolling in their coursework and,
  - software learning from their coursework.

- student focus group interviews (one per course) to explore students' discipline-specific software literacy,
- student produced work or reports as part of their learning with and through software in their coursework (where applicable in the courses investigated) to obtain an understanding of their learning outcomes and,
- ongoing informal interviews with lecturers and students as interesting themes emerge from the observations.

### 3.2.4 Limitations and Ensuring Quality of Data Collected

The data collected is based on voluntary participant participation. The findings therefore are a reflection of the extent participants were willing to be truthful and to take the time to carefully consider the questions asked. This is mitigated somewhat through the steps adopted to add rigour in the design of the different data collection strategies—survey design, interview protocol and observation protocol (see Appendix 3.2 for details).

Although the findings from the case study will not necessarily be generalisable to the wider university population, the data is sufficiently detailed to show similarities to other university contexts with similar teaching environments. By providing *rich thick descriptions* (Lincoln & Guba, 1985) of the study setting, the findings give nuanced insights into digital and software issues and practices relating to tertiary teaching and learning.

### 3.2.5 Analysis of Data

A sociocultural theoretical framework provided the overarching analytical frame to guide the data analysis (Wertsch, 1998). A sociocultural view of learning as mediated action (Wertsch, 1991a) was pertinent to this study where the focus was on how participants learned and used software as a means for accomplishing a range of goals (Cole & Engestrom, 1993; Wertsch, 1991b, 1998). This allowed the research team to understand the functioning of the individual in relation to their unique sociocultural setting and how the setting in turn influences and transforms the individual in significant ways. In this study, the focus was thus on how people-in-action are using software. The research team drew on Mietenen (2001) to view software as an artefact that carries the "intentions and norms of cognition and form part of the agency of the activity" (p. 301), and at the same time constrains a person's agency.

The purpose of the online student survey (paper versions were also made available for students' convenience) was to obtain broad brush understanding of student access to and familiarity with general software and technologies and more specifically their engagement/adoption and understanding of discipline-specific software (its affordance and constraints).[1]

The purpose of the interviews was to obtain participants lived experiences and perceptions of the discipline-specific software they were teaching/learning regarding its affordances and constraints in shaping how they teach and learn in tertiary settings. To analyse the interview data, all interviews were transcribed and imported in NVivo (version 10) qualitative software package. Each transcript was read several times. A thematic analysis based on the meaning underlying the text as opposed to the semantics of the text was then conducted. This began with the coding of key ideas

---

[1]Responses to student survey were collated within the online survey platform, *LimeSurvey*. When the survey closed, the responses captured on LimeSurvey and the paper version of the survey were entered into Microsoft Excel. Visual representations (charts) of the data were created using Excel.

within each transcript. Coding was guided in part by the sociocultural theoretical stance of the research project which directed attention to the interaction amongst and between the participants and the tools/technologies/software that they adopt to achieve productive goals. After coding each transcript, codes that were similar were either combined or sub-codes were generated. This cycle of refinement occurred throughout the analysis process. These codes were regularly revisited by the research team members and discussed for further refinement to eventually become key themes. Emergent themes were identified through a process of inductive reasoning (Braun & Clarke, 2006).

The purpose of the videotaped sessions was to obtain audiovisual/multimodal evidence for the ways lecturers were teaching the use of a particular software. To analyse the video data, all videotaped observations were imported into the NVivo 10 environment and coded. The videotaped data supplemented and triangulated the other main forms of data collection.

As part of enhancing the quality and interpretation of the data collection and interpretation, strategies, such as triangulation across multiple data sources and researchers, were employed alongside documenting an audit trail, regular team meetings and member checking of data by participants to verify their interview transcripts (Lincoln & Guba, 1985). A collaborative team approach to data analysis was further adopted to identify patterns, seek explanations for unique findings and ensure collective commitment to emergent findings and their ongoing refinement (Armstrong & Curran, 2006). Within-case and cross-case analyses were conducted to identify software literacy skills and understandings unique to and common across each discipline (media studies and engineering). Overall, the iterative and collaborative data analysis process added rigour and credibility in the research.

### 3.2.6   Participants

Details of the media studies courses and participant year levels investigated and types of data collected are shown in Appendix 3.3. Altogether four media studies courses were studied (two of which were repetitions of the same courses offered in different semesters to enable different students to complete their programme). These courses range from basic to more advanced media studies courses covering introductory media and digital practices to various levels of video production courses. The project received human ethical approval from the Faculty of Education, University of Waikato, and all participants participated on a voluntary basis.

## 3.3   Findings

The findings are reported according to the research questions (see Sect. 3.2). For each research question, quantitative data from the survey will be presented first followed by qualitative data. The quantitative data report on percentages based on the proportion of respondents' response to the survey. Representative participant quotes are provided to illustrate/evidence key themes emerging from the analyses.

### 3.3.1 To What Extent, and How Student Software Literacies Develop and Impact on the Teaching and Learning of Discipline-Specific Software

The findings for the first research question will be reported in two parts. The first part will scope the extent and how media studies students' software literacy develops (Sect. 3.3.1) while the second part reports on the ways students' software literacy development impact on the teaching and learning of software in formal tertiary settings (Sect. 3.3.2).

In order to gauge the extent students developed their software literacy skills, an understanding of their general background experience with using software and technology is warranted and will focus on understanding students' general comfort level in engaging with technology, their preference for more informal learning strategies when acquiring software skills, their understanding of core affordances and constraints of discipline-specific software applications, and evidence of critical software literacy exhibited while completing coursework.

#### 3.3.1.1 Student Comfort Level with Technologies

Media studies students' (n = 102) responses to the survey item regarding their general comfort level with engaging and adopting new technologies indicated that 37% of students reported they usually use technologies when most of their friends do (average across four papers), 36% reported liking new technologies and using them before most people they knew did, and another 20% reported loving new technologies and being among the first to use them. These results illustrate a majority of incoming students (92%) consider themselves early or quite early adopters of new technologies and are comfortable in engaging with new technologies.

#### 3.3.1.2 Student Preference for Informal Learning Strategies

Students reported they drew mostly from informal learning resources when acquiring basic skills to use media studies related software. Figure 3.1 shows findings when media studies students were asked to identify 'useful', 'very useful' and 'extremely useful' strategies for learning software.

The three highly valued learning strategies (combined 'useful', 'very useful' and 'extremely useful') by media studies students were "Going online to refer to instructions" (91%), "Asking a peer" (86%) and "Going online to refer to YouTube videos"

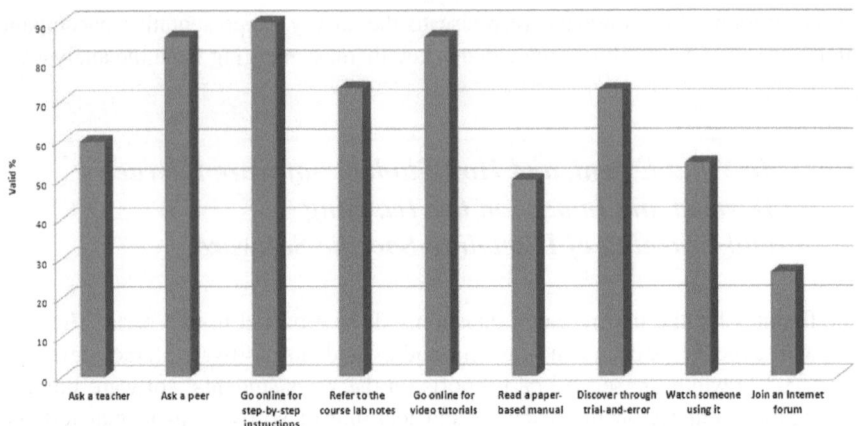

**Fig. 3.1** Strategies media studies students used to learn discipline-specific software (collated 'useful', 'very useful' and 'extremely useful'). *Note* Percentages are averaged across four papers (At the University of Waikato, the term 'papers' refer to courses). [Adapted from Khoo et al. (2016), with permission from TLRI]

(86%) as useful to their learning of discipline-based software. This trend indicates a reliance on informal learning strategies despite having formal training to learn how to use a disciplinary-specific software. Drawing from online resources and YouTube video tutorials, including asking peers and self-taught *experts* for support, were preferred over formal learning strategies. The open-ended responses in the survey affirmed these informal supports over more formal strategies.

> I would say the internet is a great, fast database for learning new things/understanding things, especially video tutorials because you can work on the software whilst watching tutorial video (Second-year media studies student).

> The main available help was from fellow students who were experts. Had they not been there (it's not their job) the work produced would have been crap as I wouldn't have known what was possible in the software. Online tutorials can be useful (Second-year media studies student).

In focus group interviews students also indicate a preference for learning at their own pace, sometimes drawing upon "more expert" peers or approaching learning collectively. Their responses most typically centred on using online materials such as YouTube instructional videos, trial-and-error and referring to a software's help feature:

> Trying to follow a software tutor in class is like watching a YouTube video without pause and rewind (Second-year media studies student).

> Or Google, anything like that and try and find people who have done it before. I think also, you know, like programmes that when you hover over the button and it comes up with what it does, are the best programmes I've ever come across because at least you can try to find something and you can do it yourself, you can actually know what every tool and thing does (Second-year media studies student).

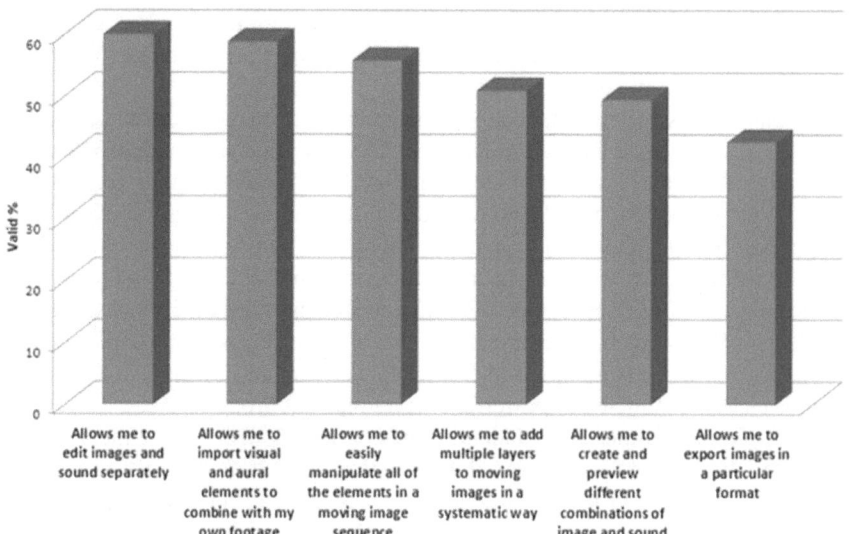

**Fig. 3.2** Student identification of the affordances of media editing software. *Note* Percentages are averaged across four papers. [Reproduced from Khoo et al. (2016), with permission from TLRI]

I find you learn how to basically do it and then you start playing around with the settings; [when you] start playing around with the settings you get your own style (Second-year media studies student).

I'd have to say ... experimentation definitely is a huge part of learning how to use it, because otherwise you don't really know what it can do. The tutorials have definitely been helpful because there's been things that you wouldn't have even thought of because you can sit there and you watch films and you sort of realise now the number of times that they do some of these things, but it just doesn't occur to you ... and learning from each other. Everyone is going to work slightly differently and they're going to come up with slightly different ideas (Second-year media studies student).

### 3.3.1.3 Student Understanding of Software Affordances and Constraints

In line with their perception of being early adopters of technology, media studies students demonstrated a familiarity with individual disciplinary software in terms of their core affordances and constraints. Figure 3.2 illustrates students' responses to a range of affordances identified in the software that they were learning to use as part of their media studies coursework and the value of these affordances for addressing tasks in their discipline.

Students indicated that media editing software such as Final Cut Pro, Adobe Creative Suite and similar applications importantly enabled them to edit images and audio files separately (60%), import audiovisual elements to combine with their own video footage (59%), easily manipulate all elements in a moving image sequence (56%) and that they could systematically add layers to a moving image (51%) when they needed to complete a creative project.

A majority of students installed their course software on their personal computing devices (across the four media studied paper surveyed) in courses using Windows™ compatible software. Students at the higher levels of university study (e.g., 84% of Year 3 Video Production students) tended to install course software on their laptop suggesting that they thought it was important to have personal copies of discipline-specific software. On the other hand, only a small percentage of first year media studies (e.g., 17% of Year 1 Video Production students) installed Final Cut Pro on their personal computers indicating they either did not view personal copies as being important in the first year, or that most students do not have Macintosh hardware at home which was sufficiently powerful to run Final Cut Pro.

Students were further able to identify the constraints within these media editing software. Their response when asked the question 'what does this software NOT let you do that you would like to be able to do?' indicated four key constraints they had encountered when using media editing software in their coursework. Student responses in the open-ended responses section of the survey alluded to limits of functionality, compatibility and difficulty in using the features of a particular software. Key examples were:

> The drawback is After Effects doesn't support the way for editing sound as Premiere (Third-year media studies student).

> It's annoying that I have to go through so many windows to get certain tools. Need more fast [shortcut] key/icons (Third-year media studies student).

When asked for suggestions to improve/enhance the use of a particular media editing software, student responses in the open-ended responses of the survey made reference to enhancing a software's functionality, including its ease of use, fixing existing issues and increasing compatibility and its accessibility. These are exemplified in the following quotes:

> Maybe to look at the technical flow of the program e.g. make things easier to find (Third-year media studies student).

> [To be able to] edit sound and real-time playing (Third-year media studies student).

> Less complicated render settings, and quicker options. Self-set macros for quickly applying desired effects (Third-year media studies student).

> Discover more plug-ins. I think Premiere/After Effects is a much better editor than Final Cut Pro (Second-year media studies student).

> After Effects = [improve on] render time. Premiere Pro = [to improve on] Real time frame rate/render time (Second-year media studies student).

> More freedom! You can create your own transitions and have a wider variety of sound manipulation (Second-year media studies student).

> Easier horizontal scrolling/clip dragging in the timeline window (Second-year media studies student).

More automated information on sequence settings when importing footage would be good (Second-year media studies student).

Making an effects panel easier to use (Second-year media studies student).

Easier integration of projects between programs (Second-year media studies student).

Students' responses overall showed they were quite knowledgeable and confident in their basic and troubleshooting skills when engaging with disciplinary based software, and quickly started to encounter the limitations of the default set-up of the application. Their ability to do so corresponds to tier 1 and 2 proficiencies of our hypothesised software literacy scheme.

### 3.3.1.4 Relative Absence of Critical Literacy Among Students

Triangulation of data sources suggested that very few media studies students report being at tier 3 of our software literacy framework (see Sect. 3.3.3 for further details and possible reasons for the lack of tier 3 observation in our data). Those who were at tier 3 proficiency level in most cases were already competent on entry to the course. The following interview excerpt with a third-year student illustrates how he had begun to develop the ability to critique software as early as high school. He had challenged his teacher's insistence on using Final Cut Pro preferring instead Adobe Creative Suite:

Like in high school we were always told to use Final Cut and every time it was Final Cut this, Final Cut that. And then there was just a point where I raised my hand to my teacher and said look, you can talk to me about Final Cut all you want but you're not going to sell me that it's better than Premiere because in my belief it's just not.

He was already proficient in terms of understanding the affordances of various software and could identify differences in the conceptual framework between different media editing software such as After Effects and Final Cut Pro:

I have very strong views about software. I just don't think that Final Cut Pro is an acceptable tool to use at a university [grade] thing ... it's a good video editing tool but it's good for stuff like documentaries where it's just your basics and stuff like that. But if you're really studying screen and media studies you should really be using After Effects because that is what a lot of the screen is about. You can't look at any films today and see absolutely no effects needed in it or anything like that.

This was but one example of students gaining tier 3 software literacy proficiency. He had taken the time, initiative and had interest to experiment and trial a range of discipline-specific software to understand their nuances to the extent he could critique their use and recommend the software that is best fit-for-purpose to achieve/complete a learning task within his discipline.

### 3.3.1.5 How the Development of Students' Software Literacy Impacts on the Teaching and Learning of Discipline-Specific Software in Formal Tertiary Setting

The development of software literacy occurred at various rates across disciplines and was strongly shaped by lecturer teaching approaches, student expectations, and disciplinary assumptions about the need to achieve professional levels of software competency. In media studies more students had prior experience using media editing software (usually at high school) and hence had a higher familiarity with the software, although the diversity in student experiences, knowledge and background meant lecturers needed to be flexible in their teaching approach. One tutor highlighted this issue of student diversity in his class:

> Because there are students that are at different levels of competency when it comes to even computer usage let alone program usage—people that are defiant that they have to learn how to use a computer or learn how to use a program, which is quite rare you would consider, like, where you're surrounded by so much technology and it requires technology to produce such work as what the class is asking (Tutor teaching second-year course).

An aspect that appeared to facilitate students' learning of discipline-specific software was students' prior engagement with artefacts or software that had a similar conceptual basis and so provided a pathway for them to engage with new and more advanced software learning; for example, media studies students with prior experience with Photoshop found it easier to pick up the skills to use other media editing software. Second-year media studies students in the Video Production course elaborated that having prior experience with similar software such as Adobe Premier (reported by 52% of students), Final Cut Pro (48%), Movie Maker (45%), After Effects (41%) and iMovie (38%) was advantageous to their current learning of software in their course. From the survey (across all four media studies papers surveyed), students highlighted four advantages of having prior experience with similar software in their learning of current course software: familiar/similar interface (reported by 31 students), transfer of skills across software (17 students), enhanced awareness of functionalities (four students) and creativity (one student). Some exemplifying quotes include:

> Almost exactly the same interface. I did not have to learn very much because I already knew how to use Final Cut Pro (Second-year media studies student).

> Even if you were not sure exactly how to do something, you know it could be done and you just need to do a google search to find out. In other words you know what to type into google to search for the feature (Second-year media studies student).

> It opened my way of thinking … I already knew how to use Photoshop before I used After Effects but I still feel using After Effects increased my understanding of Photoshop in some ways even if it was just my understanding of how the Adobe products function (Second-year media studies student).

> There are common capabilities that are not obvious, but knowing they likely exist from other software means you can find them (Third-year media studies student).

Adobe products work well together as learning Premiere helped me learn After Effects (Third-year media studies student).

It has made me more confident with experimenting with more software in the same field (Third-year media studies student).

It has not helped me understand new things, but understand old things in a new way (Third-year media studies student).

All editing software is fairly similar so I guess it became easier to navigate through new software (Second-year media studies student).

Learning other similar software helped students to learn their course software. Most students reported that a similar interface helped them negotiate the course software, and this, in turn, meant they could transfer many of the skills gained from previous software to the course software.

Students also proposed ways lecturers could approach the teaching of discipline-based software in order to enhance their appreciation of the socioculturally and historically relevant disciplinary ideas embodied within the software. In media studies, students raised the need to understand the broader contextual principles and conceptual framework behind the design of a software application for them to better appreciate its relevance and potential applications:

Like in Final Cut Pro it features words like "bins" and other words and they go back in history to, you know, actual bins that you put film footage into and the cutter will bring them out and cut them. I think that the history of editing and why those terms are used and giving them a bigger picture might just help them realise the terms. [...] it's just that deeper knowledge that's very shallow when you're coming into software if you don't know the history of the industry that goes behind it (First-year media studies student).

Importantly, students described the importance of lecturers using a range of teaching strategies to effectively cater for students' varying learning needs.

Another suggestion was to raise students' awareness of the possibilities posed within a software. Students recognised that some degree of familiarisation with software was achieved informally and discussed the value of setting some expectations of what was possible—to provide some benchmarks and motivations for their own (informal) learning.

Overall, students recommended that, in order to cater for diverse abilities, experiences and backgrounds, lecturers need to use a range of strategies (formal and informal) and to be flexible when teaching about and with software to facilitate the students' learning and development of software literacy.

### 3.3.2 Student Perception of the Software Literacies that They Learnt as Part of Their Tertiary Coursework

In relation to learning and using discipline-specific software as part of their coursework, a majority of students reported shifting in their ability to use a software at the end of a course, indicating some gains in software literacy (see Fig. 3.3).

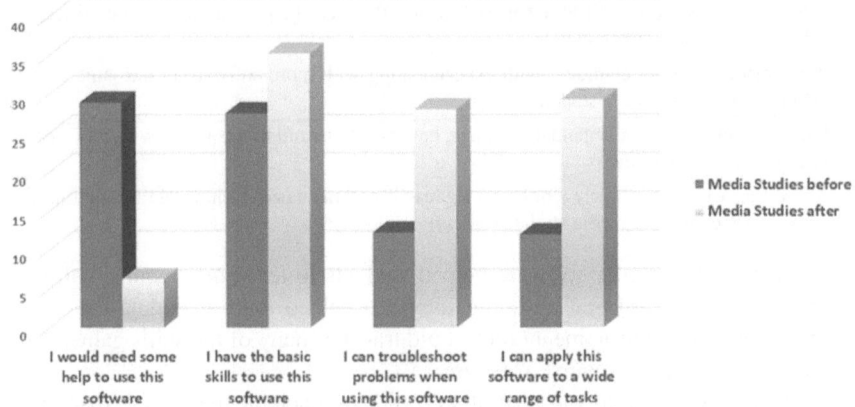

**Fig. 3.3** Changes in media studies student assessment on their ability to use discipline-specific software. *Note* Percentages are averaged across four papers. [Adapted from Khoo et al. (2016), with permission from TLRI]

When asked to rate themselves along the categories of 'I would need help', 'I have the basic skills' (level 1 of our framework), 'I can troubleshoot problems' (level 2) and 'I can apply this software' (level 3), students at the start of their coursework generally felt they would need help to use a particular media editing software (29%), or that they would only have the basic skills to use the software (28%). This decreased to 6% at the end of the course of students needing help and an increase to 35% of students who felt they now have the basic skills to use the media editing software after learning about it in the course. Another 28% of students thought they were able to troubleshoot problems faced in using the software, an increase from 12% at the beginning of the course. Gains in these two levels (basic skills and troubleshooting ability) correspond to the first two levels of our software literacy framework. By the end of the course, only 29%, however, thought they could apply their skills to a wide range of tasks, an increase from 12%, an indication of a lack in achieving the third level of our software framework. Therefore while students reported gains across all three tiers of our software literacy scheme with approximately similar gains for tiers 2 and 3 (16 and 17% respectively), the overall percentage of gains based on students achieving the second and third tier was rather low at 33%. Very few students report achieving tier 3 of our software literacy framework.

Students generally felt that developing up to tier 2 level software literacy proficiency was adequate for when they graduated from the media studies programme. In a student focus group, final year media studies students explained how tier 2 level proficiency was adequate to provide them with entry-level skills into the creative industries which they can further extend if and when needed:

> The skills and techniques of software editing stuff that I'm taking from this aren't going to be anything equivalent to what someone who's done a directed focused course in this sort of thing so I want a basic proficiency so that if I want to do some stuff on my own I could;

in the future if I have the opportunity to, I can say I can learn it, I know a bit, I can expand it; just pretty much I want a basic, broad idea of what I can do with it (Third-year media studies student).

Not like for film grade or production grade kind of stuff. Just broad skills so that I can go out into the media kind of job scenario and start at the bottom, work my way up (Third-year media studies student).

This suggests a need for lecturers to consider how they might adapt their teaching to cater to varying student diversity (those with basic to advanced software literacy skills) in the programme.

As part of understanding student perception of the software literacies that they learnt from their tertiary coursework, we were particularly interested to see the extent they understood software as influencing and shaping their disciplinary knowledge. This is unpacked in the next section.

### 3.3.2.1   How Students Understand Software as an Influence on the Way They Encounter and Make Sense of Disciplinary Knowledge

As illustrated by Fig. 3.3, only a few students achieve tier 3 of our software literacy framework. Very few students discussed or were aware of how media editing software shaped their disciplinary knowledge (a key part of software literacy). The few media studies students who did achieve tier 3 highlighted the general ways their course software helped them consider aspects of disciplinary practice such as video production. In their open response in the survey and focus group comments, tier 3 proficient media studies students commented on the freedom and versatility that their discipline-specific software provided. That is, they thought the software enabled them more freedom to create a wider range of aesthetic designs. A representative student quote revealed his understanding of the way After Effects' affordances enabled him to experiment with and come up with new ideas:

This software [After Effects] allows me to think more non-linearly, so I can place together footage/audio in new and more interesting ways (Third-year media studies student).

Other student quotes made reference to how discipline-specific software affords their working across different spaces, modalities and dimensions in new and interesting creative ways:

The ability to work in 3D space adds another dimension of possibilities (Third-year media studies student).

Gives me more confidence and freedom to manipulate my project (video sequences) in a more innovative way (Third-year media studies student).

By using the time related tools, I could control the time of the video in an interesting way, like jump or slow, or accelerate (Second-year media studies student).

The simple interface makes creating a video feel a lot easier to do which makes room for more creative risk (Second-year media studies student).

I see the shots and sounds in layers, which I slowly process through until they are refined to an overall aesthetic (Second-year media studies student).

Generally, however, a key reason for the lack of tier 3 software literacy proficiency amongst students was postulated to be due to the complexity of a particular software and the time needed to gain proficiency with using them. In focus group interviews, many students highlighted the greater investment in time and attention that the discipline-based software demanded in order to achieve basic competence (with some students explicitly commenting that they developed more intensive learning strategies in response). A media studies student considered that discipline-specific software was sufficiently complex that work done outside the classroom became an important part of learning:

> ... the [lecturers] can give you all the tools but if you're not motivated to do your own experimenting, you're not going to learn the software at all. It's all about learning something, going away, doing it, coming back, learning something, going [away], you know—it's not just a sort of teach me (Third-year media studies student).

> That's definitely a big part—having the time to actually sit down and play around with it yourself. I mean, you can't expect to just be taught the software—it's something that needs time, you've got to learn it, it doesn't happen overnight (Second-year media studies student).

His peers affirmed the time investment needed in learning software outside of class instruction:

> In order to get your head around all of the possibilities it presents you do need to do some extra homework. You do need to do some sort of external work to kind of wrap your head around everything it can do, just because it covers such a broad range of possibilities (First-year media studies student).

### 3.3.3 How and in What Ways Lecturers Model Attention to, and Use of Different Affordances in Discipline-Specific Software

When interviewed, media studies lecturers indicated they had a general understanding of the affordances and of best practice for media editing software use. They could articulate a rationale for their own practices and the relevance of these to their discipline content knowledge. They adopted a range of teaching strategies to help students grasp the affordances and relevance of a software to their coursework. However, the implementation of particular software teaching strategies was tempered by their assumptions about the level of software literacy that they felt students needed to be a work-ready graduate. One lecturer explained that he wanted his students to be able to critique a discipline-specific software (achieve tier 3 software literacy) in relation to other similar software:

> I would love it that they [students] would start [to] become very critical of the pieces of software that they're using and actually start trading ideas about what's the best thing to do, which tool to use in particular instances and ... for them to understand and I guess ultimately to throw any piece of software at them and they could start to pick it to pieces really quickly.

Another media studies lecturer felt that, by providing students with a history of software design, he could encourage them to start asking questions about the affordances of software and how these affordances fit into the bigger picture software use. Identifying the affordances of software is presumed to be fundamental to students developing the ability to critique any software in terms of its use and application:

> [The questions I want the students to be able to ask by the end of the lecture are]—how do people want to use things? How do people want to use the things that you're making? Whether they're media clips or applications or web pages, doesn't matter. What kinds of people do you think are going to be using everything you make? What can you expect them to know? What kind of affordances do you think they're going to be able to detect?

Lecturers, however, cautioned that preparing students to engage with the range of possibilities that a discipline-specific software can offer (conceptual and technical) and to apply it to other contexts (tier 3 software literacy) would require time and strategies beyond formal direct instruction in their coursework, as highlighted in the idea of performative learning in the quotes below.

> There's tutorials online, there's a whole range of stuff that they [students] can explore for themselves and that's one of the reasons why software teaching is not about being in a classroom saying "this is how you do things"; it's giving them the confidence to open up this world and explore this world on their own and … that's what we do with thinking as well in terms of this whole process. You cannot teach this notion of thinking, actually—they have to get it through practice and performance. So I see it as a performative idea of learning through action as opposed to learning through telling (Lecturer teaching third-year course).

> It's like you make connections in your brain …You learn the technicals then you go away and do something creative and then you have these two ideas in your head then you can link them. Like everything just seems like to come together. It's just natural that you find the two things together then you think 'oh that might work'… takes exploration because you go out and 'Oh this doesn't seem as great as I thought, how do you modify it to be better' (Lecturer teaching second-year course).

Finally, a first year course tutor describes his approach in terms of encouraging students to model their efforts on "exceptionally great" pieces of work as one approach to students' learning and becoming aware of a software's affordances to harness its technical and creative possibilities:

> That is all done by example, so looking through people that have done exceptionally great pieces of work that exemplify sound, lighting, camera work, cinematography, and talking to them about if you apply yourself you can produce the same type of things; maybe not with the equipment that is provided but these are what we want you to try and think of in your mind and try and maybe conceptualise and actually produce.

These different expectations and assumptions were played out in the teaching approach to discipline-specific software learning; for example, observations from the formal lab sessions revealed that some tutors spent more time teaching the basic functions and affordances of a software while other lecturers focused on higher level affordances such as teaching the tasks that can be performed within a software, point out the complexities of the affordances, and how these affordances could potentially enable and constrain creative practice.

To sum up, lecturers understood that they could not provide fully immersive training in the applications they introduced at tertiary level. They sought instead to ensure that students had appropriate learning strategies to empower them to understand how to begin to apply those strategies to different contexts and to explore the software on their own terms.

## 3.4   Summary

The findings from this case study of media studies students' software literacy development highlighted students acknowledgement of being early adopters of technology, knowledgeable in the affordances and constraints of their disciplinary software, and preferred informal learning strategies to supplement their formal learning of disciplinary software. Students lacked a critical awareness of the role of software in shaping their learning of disciplinary knowledge. Factors such as the complexity of a particular discipline-based software and the time needed to develop proficiency were perceived to be key hindrances to achieving level 3 software literacy proficiency. Students who enter the media studies programme, however, came from diverse backgrounds with varying skills (some already proficient while others less so), and expectations (with a majority assuming a level 2 software literacy level to be adequate as an entry level qualification into the creative industry). The few students who achieved level 3 proficiency were able to critique the uses of a range of discipline specific-software and effectively tap into an appropriate software's affordances to extend their creative and critical abilities. The findings and their implications will be revisited and discussed further in Chap. 5.

## Appendices

## Appendix 3.1: Software Literacy Survey for Media Studies Students

Dear students,

We are interested in your learning experiences and opinions of **video editing software** packages as part of your university coursework so that we can improve students' learning experiences. Your participation is voluntary and will not impact on your course grade in any way. Your answers will be kept confidential. Your lecturer will not know the identity of students who participated in this survey.

There are 20 questions which should take you approximately 20 min to complete.

Please answer ALL questions. By answering this survey, you give your informed consent to participate in this survey.

Thank you.

Elaine Khoo, Craig Hight, Rob Torrens, Bronwen Cowie

Research Team

**Section 1. Your background information**

1. **Please indicate your age group.**
   Please choose **only one** of the following:

   ☐ Under 18
   ☐ 18–21
   ☐ 22–25
   ☐ 26–30
   ☐ 31–35
   ☐ 36–40
   ☐ 41–45
   ☐ 46–50
   ☐ Over 50

2. **Please indicate your gender.**
   Please choose **only one** of the following:

   ☐ Female
   ☐ Male

3. **Are you a domestic or international student?**
   Please choose **only one** of the following:

   ☐ Domestic student
   ☐ International student

4. **What is your first language?**
   Please choose **only one** of the following:

   ☐ English
   ☐ Māori
   ☐ Other: _____

**Section 2: Your software experience before attending this paper**

We would like to know more about your experience with software before attending this paper.

5. **Which of the following best describes you?**
   Please choose **only one** of the following:

   ☐ I love new technologies and am among the first to experiment with and use them
   ☐ I like new technologies and use them before most people I know
   ☐ I usually use new technologies when most people I know do
   ☐ I am usually one of the last people I know to use new technologies
   ☐ I am skeptical of new technologies and use them only when I have to

6. **What software have you used for video editing before attending this course?**
Please choose **all** that apply:

☐ Final Cut Pro
☐ Adobe Premiere
☐ Movie Maker
☐ iMovie
☐ Other (please tell us the name of the software): _____

7. **Which software are you are using for video editing in this course that you would consider yourself to be the most skilled?**
Please choose **only one** of the following:

☐ Final Cut Pro
☐ Adobe Premiere
☐ Movie Maker
☐ iMovie
☐ Other (please tell us the name of the software): _____

**For the next part of this survey, please use the software you had selected in Question 7.**

8. **Thinking back, how good were you in using this software before enrolling in this paper.**
Please choose only one of the following:

☐ I would have needed some help to use this software
☐ I had the basic skills to use this software
☐ I could troubleshoot problems when using this software
☐ I could apply this software to a wide range of tasks

**Section 3. Your software learning from this paper**

We are interested to know what helped you in your learning of software as part of your coursework.

9. **Please choose which of the following strategies were useful to your learning of this software.**
Please choose **all** that apply:

☐ Ask the course lecturer/tutor/an expert
☐ Ask a friend/peer/senior student
☐ Refer to the course lab notes
☐ Read a paper-based manual/step-by-step instruction booklet
☐ Go online/refer to the Internet for step-by-step instructions
☐ Go online/refer to the Internet for video tutorials (e.g., YouTube) to watch how to use it
☐ Watch someone using it in a face-to-face (physical) setting (not through videos)

    ☐ Discover through trial-and-error/practise
    ☐ Join an Internet forum (e.g., a discussion forum to ask other users for help)

10. **Please tell us what other <u>additional learning strategies</u> you had used (if any) to be able to use this software in this paper?**

    _____

11. **What are the THREE <u>most</u> useful capabilities of this software in helping you to put together a video project?**
Please choose only **three** of the following:

    ☐ Allows me to import visual and aural elements to combine with my own footage
    ☐ Allows me to edit images and sound separately
    ☐ Allows me to add multiple layers to moving images in a systematic way
    ☐ Allows me to easily manipulate all of the elements in a moving image sequence
    ☐ Allows me to create and preview different combinations of image and sound
    ☐ Allows me to export images in a particular format
    ☐ Other: _____

12. **Please tell us how this software supports you to think differently about constructing a video:**

    _____
    _____

13. **Did you install this software on your own computer/laptop?**
Please choose **only one** of the following:

    ☐ Yes
    ☐ No

14. **After learning and using this software in this course, how good would you rate yourself at using it?**
Please choose **only one** of the following:

    ☐ I would need some help to use this software
    ☐ I have the basic skills to use this software
    ☐ I can troubleshoot problems when using this software
    ☐ I can apply this software to a wide range of tasks

15. **How has learning the software helped you in your learning to use other software as a media studies student?**

    _____
    _____

16. **What does this software <u>NOT</u> let you do that you would like to be able to do?**

    _____
    _____

17. **Have you encountered any unexpected issues when using this software?**
Please choose **only one** of the following:

☐ Yes (please describe the main issue): _____
☐ No

18. **Any suggestions on how to improve this software if you had the opportunity to?**

_____

_____

19. **Have you used any other software that you consider similar to this software?**
Please choose **only one** of the following:

☐ Yes (please tell us the name of the software): _____
☐ No

20. **If you said 'yes' in question 19, in what ways are the two software packages similar:**

_____

_____

**Thank you for your time and help!**

## Appendix 3.2: Ensuring Quality of Data Collected

For the survey design:

1. Questions were crafted by referring to past literature (see for example, Dahlstrom, 2012; Hegarty et al., 2010; Massachusetts Department of Elementary and Secondary Education, 2013; Pagram & Cooper, 2011; Shih & Chuang, 2013) with the research question and intentions in mind,
2. All key terms were defined to clarify their meaning in the survey,
3. The survey underwent different cycles of item refinement where the items were debated and further refined through regular team member meetings and conversations (a form of member checking),
4. The survey was subjected to a pilot study with 26 volunteer students who were not part of the research to enhance the accuracy, clarity of questions, reduce misinterpretation and any cultural bias. Using a combination of closed-likert items, ranking type questions, and open-ended questions in the survey allowed for more detailed individual responses,
5. The survey was constructed using LimeSurvey (a free online tool) and made available online for a period of time to ensure students could access it at their convenience. Additionally, a paper-based version of the survey was also made available to students should they prefer to complete a hard copy, and,
6. The survey results were triangulated with other forms of data collection such as interviews and observations.

For the interview protocols:

1. Questions were constructed based on the research questions and by referring to sample questions asked in the literature and refined through several series of researcher meetings,
2. The interview questions were forwarded to the participants beforehand so they could prepare/consider them more carefully before the interview session,
3. Notes were taken during the interviews to document key ideas in the conversation and as a reference point (to form an audit trail),
4. Each interview was transcribed and unclear points were re-checked with the audio recordings,
5. Triangulation of the interview data with other forms of data collection was conducted to form a detailed case study of each disciplinary programme.

For the observation protocols:

1. Field notes were taken during the observations that can inform the research questions (for audit trialling), and,
2. Debriefs (post observation interviews) with the lecturers allowed the research team opportunities for further clarification and understanding of particular lecturers' motives/actions during the observation (a form of member checking).

## Appendix 3.3: Details of the Media Studies Courses Investigated and Types of Data Collected

| Courses surveyed in the first year of the study | Data collected from the different participant groups |
| --- | --- |
| Media and digital practices (Year 2, Media Studies students) A second-year introductory course on critical and creative perspectives shaping digital media practice Formal learning of creative software (e.g. After Effects, Photoshop, Premiere Pro, Illustrator) as examples of digital media practice was through tutorials, lab-based learning followed by a group project to create a digital media project | The course had 34 students Data were collected from: – 25 student surveys – 6 student final assignments on Software Literacy – pre- and post-lecturer interviews – lab observations of student learning of creative software – class observations of theoretical concepts teaching and learning, and – a student focus group interview with 4 students |

| | |
|---|---|
| Video Production Level 1<br>(Year 1, taken by Media Studies students from different learning option papers)<br>A first year introductory paper on the theory and practice of image production. Formal learning of software (e.g. Final Cut Pro and Adobe Premiere) is conducted in labs and through individual projects with peer and lecturer feedback | Data were collected from:<br>– 24 student surveys<br>– pre- and post-lecturer interviews<br>– lab observations of student learning of software—Final Cut Pro<br>– tutor focus group interview attended by 4 tutors, and<br>– an individual student interview |
| *Courses surveyed in the second year of the study* | *Data collected from the different participant groups* |
| Video Production Level 2<br>(Year 2, Media Studies students, offered in Semester A)<br>A second-year course covering advanced practical and critical understanding of the video production process in order to become more reflective as creative practitioners. Formal learning of discipline specific software is through student group work to write, produce, direct and edit a short film | Data were collected from:<br>– survey completed by 9 students<br>– a tutor individual interview, and<br>– a focus group interview attended by 9 students |
| Video Production Level 3<br>(Year 3, Media Studies students, offered in Semester A)<br>A third year paper aimed at developing students' critical thinking about their own creative practice through the production and post-production of digital video projects<br>Formal learning of software is through workshops and use of discipline specific software such as After Effects, Garage Band, Sound Track Pro and Studio Pro for post-production purposes as students develop, refine and produce individual films | Data were collected from:<br>– a survey completed by 6 students, and<br>– a focus group interview with 6 volunteer students |
| Video Production Level 2<br>(offered in Semester B) | Data were collected from:<br>– a survey of 20 students<br>– lecturer interview, and<br>– a focus group interview with 14 students |
| Video Production Level 3<br>(offered in Semester B) | Data were collected from:<br>– a survey of 19 students<br>– lecturer interview<br>– tutor interview<br>– class observation of student presentation of the project work, and<br>– individual student interviews with 2 students |

# References

Armstrong, V., & Curran, S. (2006). Developing a collaborative model of research using digital video. *Computers & Education, 46*(3), 336–347. doi:10.1016/j.compedu.2005.11.015.

Arzi, H. J. (1988). From short-to long-term: Studying science education longitudinally. *Studies in Science Education, 15*(1), 17–53. doi:10.1080/03057268808559947.

Bell, P. (2004). On the theoretical breadth of design-based research in education. *Educational Psychologist, 39*(4), 243–253.

Braun, V., & Clarke, V. (2006). Using thematic analysis in psychology. *Qualitative Research in Psychology, 3*(2), 77–101.

Cole, M., & Engestrom, Y. (1993). A cultural-historical approach to distributed cognition. In G. Salomon (Ed.), *Distributed cognitions: Psychological and educational considerations* (pp. 1–46). New York, NY: Cambridge University Press.

Dahlstrom, E. (2012). *ECAR National Study of Undergraduate Students and Technology, 2012.* Educause Center for Applied Research. Retrieved from http://net.educause.edu/ir/library/pdf/ERS1208/ESI1208.pdf.

Gall, M. D., Borg, W. R., & Gall, J. P. (1996). *Educational research: An introduction.* White Plains, NY: Longman.

Gilbert, J. (2005). *Catching the knowledge wave? The knowledge society and the future of education.* Wellington, NZ: NZCER Press.

Hegarty, B., Penman, M., Kelly, O., Jeffrey, L., Coburn, D., & McDonald, J. (2010). *Digital information literacy: Supported development of capability in tertiary environments.* Wellington, New Zealand: Ministry of Education. Retrieved from http://www.educationcounts.govt.nz/publications/tertiary_education/80624.

Khoo, E., Hight, C., Torrens, R., & Cowie, B. (2016). *Copy, cut and paste: How does this shape what we know?* Final report. Wellington: Teaching and Learning Research Initiative. Retrieved from http://www.tlri.org.nz/tlri-research/research-completed/post-school-sector/copy-cut-and-paste-how-does-shape-what-we-know.

Lincoln, Y. S., & Guba, E. (1985). *Naturalistic inquiry.* Beverly Hills, CA: Sage.

Livingstone, S., Wijnen, C. W., Papaioannou, T., Costa, C., & del Mar Grandío, M. (2014). Situating media literacy in the changing media environment: Critical insights from European research on audiences. In N. Carpentier, K. C. Schröder, & L. Hallet (Eds.), *Audience transformations: Shifting audience positions in late modernity* (Vol. 1, pp. 210–227). Routledge, NY: Routledge Studies in European Communication Research and Education.

Manovich, L. (2006). After effects or the velvet revolution. *Millennium Film Journal, 45*(46), 5–19.

Massachusetts Department of Elementary and Secondary Education. (2013). *Technology Self-Assessment Tool (TSAT).* Retrieved from https://www.surveymonkey.com/r/BGMFNF8.

Maykut, P., & Morehouse, R. (1994). *Beginning qualitative research: A philosophic and practical guide.* London, UK: Falmer.

Mietenen, R. (2001). Artifact mediation in Dewery and in cultural-historical activity theory. *Mind, Culture, and Activity, 8,* 297–308.

Pagram, J., & Cooper, M. (2011). E-yearning: An examination of the use and preferences of students using online learning materials. In T. Hirashima, et al. (Eds.), *Proceedings of the 19th International Conference on Computers in Education. Chiang Mai, Thailand,* (pp. 712–716). Retrieved from https://www.nectec.or.th/icce2011/program/proceedings/pdf/C6_S18_163S.pdf.

Patton, M. Q. (2002). *Qualitative research and evaluation methods* (3rd ed.). Thousand Oaks, CA: Sage Publications.

Selwyn, N. (2010). Degrees of digital division: Reconsidering digital inequalities and contemporary higher education. RU&SC. *Revista de Universidad y Sociedad del Conocimiento, 7*(1), 33–42. Available at http://redalyc.uaemex.mx/src/inicio/ArtPdfRed.jsp?iCve=78012953011.

Shih, C.-L., & Chuang, H.-H. (2013). The development and validation of an instrument for assessing college students' perceptions of faculty knowledge in technology-supported class environments. *Computers & Education, 63,* 109–118. doi:10.1016/j.compedu.2012.11.021.

Wertsch, J. V. (1991a). *Voices of the mind: A sociocultural approach to mediated action*. Cambridge, MA: Harvard University Press.

Wertsch, J. V. (1991b). A sociocultural approach to socially shared cognition. In L. B. Resnick, J. M. Levine, & S. D. Teasley (Eds.), *Perspectives on socially shared cognition* (pp. 85–100). Washington, DC: American Psychological Association.

Wertsch, J. (1998). *Mind as action*. New York, NY: Oxford University Press.

Yang, X. (2014). Teaching and learning fused through digital technologies: Activating the power of the crowd in a university classroom setting. In D. J. Loveless, B. Griffith, M. E. Berci, E. Ortlieb, & P. M. Sulivan (Eds.), *Academic knowledge construction and multimodal curriculum development* (pp. 77–85). Hershey, PA: IGI Global.

# Chapter 4
# The Learning, Use and Critical Understanding of Software in Engineering

**Abstract** This chapter (as with Chap. 3) details the findings from a two-year funded empirical study aimed at understanding tertiary students' development of the understandings and skills needed to use software as forms of software literacy. Two case studies were developed. A case study of engineering students' software literacy development is the focus of this chapter. Two cohorts of students were tracked using mixed methods to explore their learning and understanding of discipline-specific software (here the Computer-Aided Design (CAD) software SolidWorks). An additional group of advanced final year CAD students were also interviewed to ascertain if there were particular nuances in their software learning experience. The findings of this case study provide insight into engineering students' software literacy development in a specific tertiary context. A discussion of the findings including implications for what the findings might mean in relation to the wider field of software teaching and learning is addressed in Chap. 5.

## 4.1 Introduction

Engineering students are expected to have knowledge and be proficient in discipline-specific software as part of their learning and becoming a professional engineer. International engineering professional accreditation agreements, such as the Washington Accord (2013), detail the broad range of graduate attributes and professional competencies that today's engineering graduates need. The Accord states that the fundamental purpose of engineering education is to build each graduate's knowledge base and attributes so they can continue learning and developing the competencies required for independent practice beyond formal learning contexts. In engineering, students must be able to visualise and rotate objects in three-dimensional (3D) space and to pictorially represent complex ideas. Part of the expectation is for students to develop some degree of discipline-based software competency to communicate their ideas clearly in order to remain competitive, and contribute productively to 21st century engineering workplaces where software-supported engineering design, process and workflow are integral components. There is evidence that different digital technologies can significantly shape how and what millennial engineers

© The Author(s) 2017
E. Khoo et al., *Software Literacy*, SpringerBriefs in Education,
https://doi.org/10.1007/978-981-10-7059-4_4

learn (Johri, Teo, Lo, Dufour, & Schram, 2014). This has not been investigated in the New Zealand context. As detailed in Chap. 1, the notion of software literacy is a potentially useful framework to understand the ways engineering students come to understand and develop their software literacy proficiencies.

This chapter, as with Chap. 3, aims to explore, examine and theorise on how the notion of software literacy is understood, developed and applied in tertiary teaching and learning contexts, and the extent to which this understanding is useful when translated into new contexts of learning with and through software. This understanding is crucial and relevant to ensure all students and lecturers are better supported in teaching and learning processes that are mediated through and focused on software. Sociocultural theoretical perspectives are adopted as a basis of our study. The chapter begins by describing the study context—a case study of the experiences of undergraduate engineering students, and briefly revisits key research design ideas adopted in the study (see Chap. 3, Sect. 3.2 for details of and rationale for the research design, data collection methods and theoretical framing used in the analysis of the data). The findings from the data are detailed next to evidence in the ways students engage with discipline-specific software.

## 4.2  Research Design and Context

This research draws from findings from our two-year longitudinal research project funded by the New Zealand Ministry of Education: Copy, cut and paste (CCP): How does this shape what we know? (Khoo, Hight, Torrens, & Cowie, 2016), to report on the views of participating tertiary media studies and engineering students from a New Zealand university. The research questions framing the investigation of the case study were:

1. To what extent, and how does student software literacy develop and impact on the teaching and learning of discipline-specific software in formal tertiary teaching settings?
2. What software literacy do students consider they learnt as part of the case study tertiary course(s)?
3. How and in what ways do lecturers model attention to and use of different aspects of software affordance in a course which utilises discipline-specific software?

The study was underpinned by a qualitative interpretive methodology (Bell, 2004; Maykut & Morehouse, 1994) which is consistent with a sociocultural perspective in valuing the social and cultural contexts for how knowledge is co-constructed through interaction between individuals and tools (Wertsch, 1998). As with the media studies case reported in Chap. 3, a case study of engineering students and lecturers was adopted (see Sect. 3.2 for details). The research team collaborated with lecturers who were keen to examine the notion of software literacy through the teaching and learning of commonplace discipline-specific software—SolidWorks© (a computer-aided design (CAD) software that enables 3D drawings and is highly regarded in engineering industries). The engineering case study was located within the under-graduate mechanical engineering programme at the University of Waikato, New Zealand.

### 4.2.1   The Engineering Case

The engineering programme is characterised by high enrolments of students with diverse backgrounds (generally young school leavers with a small proportion of international students) at entry level (about 180 students). Students attend lectures and engage in the design principles and process through examining and discussing case studies of designs. They also attend supervised laboratory-based training where they are provided with tasks to help them acquire further proficiency with SolidWorks and work on individual assignments. Students are required to participate in group design projects as a demonstration of their SolidWorks-supported design understanding and application. CAD software, such as SolidWorks, is considered an integral component of modern engineering and is widely used in industry. No familiarity with CAD or drawing software is assumed for entry into university coursework, although students are expected to be familiar with the use of computers.

All four-year engineering degrees in New Zealand require the completion of 800 h of appropriate workplace experience. Not all work placements will include the use of CAD; however, for those that do, it is useful to consider how students transition or adapt their learning (and learning strategies) from the tertiary environment to the particular demands of their workplace, including learning alternative CAD appli-cations. Knowledge of CAD can still be useful for students not actively using the software to allow them to interpret CAD generated drawings and usefully contribute to design discussions.

### 4.2.2   The Research Design

Adopting the overlapping longitudinal study design (Arzi, 1988) in the context of the engineering case study enabled the research team to map student learning and development across the entire four years of the engineering degree programme in two years. Within the engineering case, a range of engineering courses focusing on the teaching and application of SolidWorks were investigated. The research team tracked:

- One group of students from Year 2 of engineering design coursework into Year 3 coursework,
- A smaller group of Year 2 students into their work placement to study their ability to transfer and apply or adapt their SolidWorks software literacy in the more immersive and/or specialised forms of practice required within workplace settings.

The extent engineering students are aware of and can apply their discipline-specific software literacy in the workplace context and how this, in turn, shapes their understanding of their disciplinary knowledge is of interest in this study, and

- A separate advanced group of elite Year 4 students selected to represent the university at a prestigious Formula SAE-A competition highly regarded by the industry and considered to have sophisticated software literacy skills. Each elite team must design, build and race a small high-performance race car.

The overall assumption is that students' software literacy develops as they gain more experience with the SolidWorks software although a linear progression was not assumed.

### 4.2.3  Data Collection

As with the media studies case, multiple data were collected to address the research questions through:

- lecturer individual interviews and tutor focus group interviews (one per course) to obtain lecturer/tutor perceptions and awareness of the affordances of the Solid-Works software and how this influenced the teaching and learning of the software,
- observations of lectures and laboratory (lab) sessions (up to two observations per course) to understand students' learning to use SolidWorks,
- online student surveys to obtain student evaluation on the teaching and learning of SolidWorks at the end of each course (see Appendix 4.1). The survey consisted of four key sections asking students about their:

  - demographic profile,
  - SolidWorks experience prior to enrolling in their course,
  - SolidWorks learning from their course, and
  - evaluation of SolidWorks.

- student focus group interviews (one per course) to explore students' SolidWorks software literacy,
- student produced work placement reports as part of their learning and application of SolidWorks during their work placements, and
- ongoing informal interviews with lecturers and students as interesting themes emerged from the observations.

The data collected were based on voluntary participant participation. The findings therefore are a reflection of the extent participants were willing to be truthful and to take the time to carefully consider the questions asked.[1]

---

[1]Readers are referred to Sects. 3.2.4 and 3.2.5 for a discussion on the limitation of the study, ways of enhancing the study's rigour, the rationale for the sociocultural perspective adopted in the analysis of the data and the analytical processes undertaken to enhancing the quality and interpretation of the

### *4.2.4 Participants*

Details of the engineering courses investigated, participant year levels and types of data collected are shown in Appendix 4.2. Altogether three engineering courses were investigated with an additional focus group interview conducted with a group of selected elite final year students. These courses cover increasingly sophisticated ideas related to engineering design and processes and require compulsory student learning of the SolidWorks software. The project received human ethical approval from the Faculty of Education, University of Waikato, and all participants participated on a voluntary basis.

## 4.3 Findings

The findings are reported according to the research questions. For each research question, quantitative data from the survey will be presented first followed by qualitative data. The quantitative data report on percentages based on the proportion of respondents' response to the survey. Representative participant quotes are provided to evidence key themes emerging from the analyses.

### *4.3.1 To What Extent, and How Student Software Literacy Develops and Impacts on the Teaching and Learning of Discipline-Specific Software*

The findings for the first research question are reported in two parts. The first part scopes the extent and how engineering students' software literacy develops (Sect. 4.3.1) while the second reports on the ways students' software literacy development impacts on the teaching and learning of software in formal tertiary settings (Sect. 4.3.2).

In order to gauge the extent students developed their software literacy skills, an understanding of their general background experience with using software and technology is warranted and will thus focus on understanding students' general comfort level in engaging with technology, their preferred learning strategies when acquiring software skills, their understanding of core affordances and constraints of individual software applications, and evidence of critical software literacy exhibited while completing coursework.

---

data collected. As with the media studies case, within-case and cross-case analyses were conducted to identify software literacy skills and understandings unique to and common across each discipline.

#### 4.3.1.1    Student Comfort Level with Technologies

Engineering students' comfort level in terms of engaging and adopting new technologies were elicited. Sixty-seven second-year engineering students out of a class of 140 students responded to the survey. When asked about their general views towards adopting new technologies, 43% of students indicated they usually use new technologies when most of their friends do, 31% reported liking new technologies and using them before most people they know do, and another 10% indicated they love engaging with new technologies and are among the early adopters to use them. These results highlight, therefore, that a majority of students (84%) consider themselves early or quite early adopters of new technologies and are comfortable in engaging with new technologies.

Next, we report on the strategies and resources the Years 2–4 students described as supporting their learning and use of SolidWorks.

#### 4.3.1.2    Students' Preferred Learning Strategies

When asked about their preferred learning approaches when acquiring basic skills to use SolidWorks, students' responses tended to favour a combination of formal and informal learning strategies (see Fig. 4.1).

The three highly valued learning strategies (combined 'useful', 'very useful' and 'extremely useful') by engineering students were "Asking the course lecturer" (80%), "Asking a peer or self-taught experts" (49%), and "Refer to the course/lab notes" (41%). Other strategies included "Read a paper-based manual" (35%), "Go online for step-by-step instructions" (30%), "Go online for video tutorials" (25%), "Watch

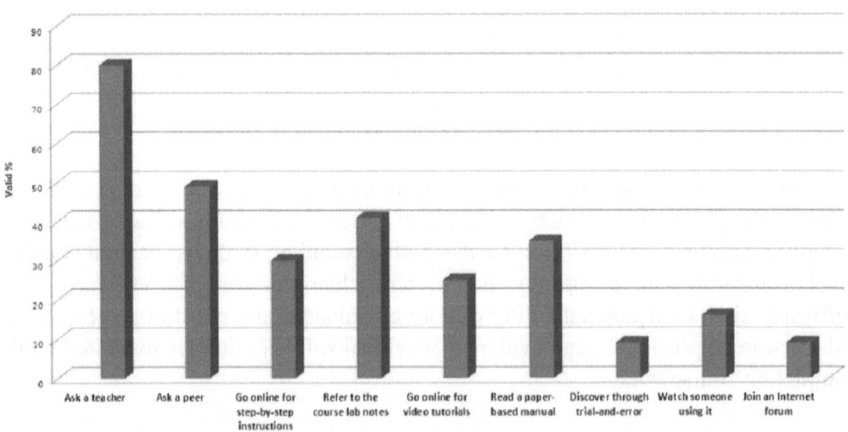

**Fig. 4.1**  Strategies engineering students used to learn SolidWorks (collated 'useful', 'very useful' and 'extremely useful'). *Note* Percentages are averaged across four papers (At the University of Waikato, the term 'papers' refer to courses). [Adapted from Khoo et al. (2016), with permission from TLRI]

someone using it" (16%) and so forth. Overall, apart from asking the course lecturer, the reported strategies tend to draw from more informal resources and own initiatives that occurred outside of course or lab hours. Possible reasons for engineering students valuing asking their lecturer for help before relying on more informal strategies as compared to media studies students could be due to the perceived complexity of SolidWorks or less exposure/experience at high school with CAD software in general.

In the open-ended responses of the survey and focus groups, students elaborated on drawing from a variety of resources to help them learn to use SolidWorks. They recognised that the conceptual and technical complexity of SolidWorks demands a more self-directed and committed investment in time to learn the software, which required developing informal learning strategies to complement the formal training provided within their tertiary programme. This led students to draw from SolidWorks' in-built tutorials, 'more expert' peers, practise through using their intuition and trial-and-error, as well as using online materials such as YouTube instruction videos (which notably involved developing an expertise in finding instructional material suited to "their level"). In fact, 76% of participants reported installing SolidWorks on their personal computers or laptops in order to be able to practise and use the software for their coursework.

> I learnt heaps just by having SolidWorks at home and then just grabbing something and going, 'Oh well, I wonder if I can model this?' And I think everyone learns quite well like that because then if you're talking to people and you would be like, 'I modelled this watch' and they're going, 'That's cool, how did you do that?' And then you can just talk through it and go, 'Oh, that's a neat way of doing it.' (Fourth-year engineering student).

> I've been working next to a fourth-year I'm friends with and he's looked at my work and gone, 'Whoa, dude, hold on—let me show you how to do this' and he's stepped in and shown me a whole bunch of stuff (Second-year engineering student).

> Most of my learning on SolidWorks has been done by working on it at home or playing around at home, e.g., how to do that, learning from peers and also YouTube videos. Like, if there's no one around and you can't do it, type it into Google, type it into YouTube and hopefully you'll get something and if you don't then get some help (Fourth-year engineering student).

> You're never going to learn it just by sitting in a class and having someone preach to you … Because it gets frustrating sometimes, but once someone teaches you the basics of sketches and you learn those things and then you can start experimenting and troubleshooting and stuff and then using the different features [in SolidWorks] and that gets you nice and efficient (Fourth-year engineering student).

This practice of mainly drawing from informal learning strategies continued when students were in their work placement. For example, learning from peers was common informal workplace learning practice which added to students' software literacy development:

> I know that in my work placement, I had a couple of people who knew how to do everything and I would ask them. There was some stuff that they didn't know and there were some things that I'd learnt at uni that they didn't know existed in SolidWorks (Third-year engineering student).

Another student affirmed the value of this strategy when thrown into a challenging real world context to use the software appropriately:

> On my first day I think I was sat down and he was like, 'Right, make this' and I made it and he was like that's totally wrong and then spent like three days teaching me how to use it, just how he liked it taught so (Third-year engineering student).

Other students found working alongside engineering professionals offered them authentic learning experiences to apply their software skills in relevant ways:

> And then probably for further development [learning of SolidWorks] was dealing with industry professionals—working with them, getting them to critique my modelling for some stuff that I was having manufactured because it's massive and I learnt a lot about stainless design (Fourth-year engineering student).

This combination of formal and informal learning strategies adopted in the formal university context and in real world work placement contexts contribute to students developing software literacy skills and confidence.

#### 4.3.1.3  Student Understanding of Software Affordances and Constraints

In line with their perception of being early adopters of technology, engineering students demonstrated a basic familiarity with SolidWorks and easily identified its key affordances and constraints. For example, when asked their views on how Solid-Works affords their addressing of engineering design issues, students indicated it allowed them to rotate and manipulate different views of their drawings (81%), to easily modify their drawings (79%), to draw an object to see what it looks like (or to share with others my drawing so they know what I mean) (78%) and to design and draw things before building them (see Fig. 4.2). Their ability to discern the general affordances of the SolidWorks software correspond with tiers 1–2 of our hypothesised software literacy framework including identifying the value of these affordances for addressing engineering design tasks.

Additional student views in the open-ended responses in the survey alluded to the ways SolidWorks facilitates being able to "communicate my ideas" and to specify "properties of objects required, e.g., volume, weight" for further exploration.

Students were further able to identify the constraints/limitations within Solid-Works. Their response when asked the question 'what does this software NOT let you do that you would like to be able to do?' indicated three key constraints they had encountered when using SolidWorks in their coursework. Over 28% of participants identified these in the open-ended survey responses in relation to limits with:

1. accessing the software (e.g. affordability, unable to install on their personal laptops, incompatibility in opening saved files on other computers, software crashing often). Examples of students' experience were:

> Save files: Once I found that some drawing files that I had completed at home couldn't be opened elsewhere.
>
> It runs slow and crashes often.

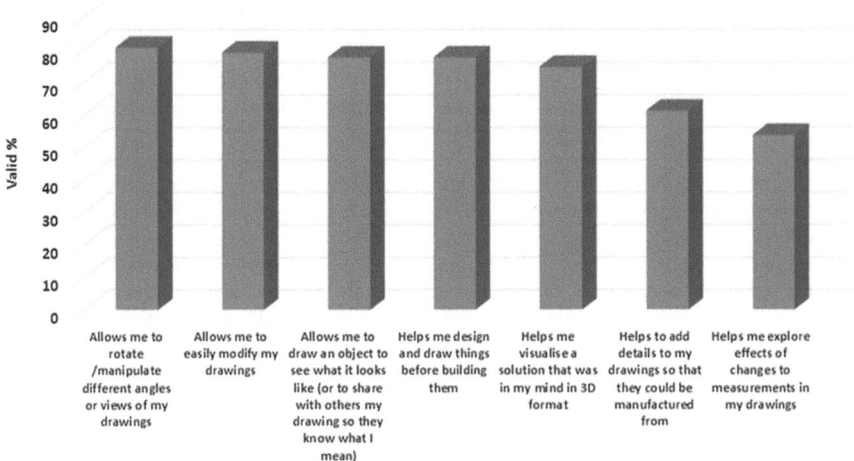

**Fig. 4.2** Student identification of SolidWorks' affordances in addressing engineering design tasks. [Reproduced from Khoo et al. (2016), with permission from TLRI]

2. learning to use the software. A student quote included:

> SolidWorks has a learning curve which can make things harder to do.

3. using particular features of the software, such as:

> The menus are too large, it makes it hard to find the tool you are looking for.
>
> To be able to move my gears after I add one assembly file to another.
>
> When using a circular pattern (during a sketch), the pattern does not copy relations i.e. being coincident on a circle.
>
> Changing 2D to 3D drawing.

Students offered suggestions to improve/enhance the use of SolidWorks. In focus groups and the open-ended responses of the survey, they suggested strategies such as having more technical guidance (e.g., "The programme crashes every hour, need to be told how to reduce computer requirements"), more in-built user-friendly support within the software (e.g., "I would like more 3D modelling support to be implemented", and, "Too many methods to do same task") and enhanced functionality (e.g., "Applying/bringing real images into SolidWorks").

#### 4.3.1.4   Relative Absence of Critical Literacy Among Students

Triangulation of data sources suggested that very few engineering students report being at tier 3 of our software literacy framework.[2] The few students who were at

---

[2]Note: Also refer to Sect. 4.3.3 for further details and possible reasons for the lack of tier 3 observation in our data.

tier 3 proficiency level in most cases were already competent on entry to the course and sought to continue extending their proficiency level as acknowledged by this student. He viewed his coursework as providing him with an initial platform to learn some basic skills which will need further extending through his taking initiative to do so:

> Like, I used AutoCAD a bit before I came to university but not with any depth and it's just playing around with things that I was making at home. And then SolidWorks, I did a lot of when we were doing the basic course. I played with a lot of different ideas in Solid Works then, not just draw boats and stuff. And then from that I wanted to put it in a practical sense and got a design job. At that time we were actually using AutoCAD but I'd bring both softwares together and sort of built the model in SolidWorks and brought it into AutoCAD and do other manipulation in AutoCAD. If you like SolidWorks you'll pick it up and use it in other things, not just at university. But [uni] will give you the base, the essential knowledge to start using it (Third-year engineering student).

In focus groups, students confirmed that SolidWorks was a complicated discipline-specific software which most were not familiar with prior to tertiary study:

> It's entry level for us, it's [SolidWorks is] super complicated—we should be learning it at that lower level, then advancing later on; or it should be a paper unto itself, because it makes the rest of the paper so difficult (Second-year engineering student).

As a result, students were less likely to identify themselves as 'highly proficient' or 'expert' in using SolidWorks at the completion of their course. Most, nevertheless, reported confidence in being able to troubleshoot issues faced when using SolidWorks (tier 2 software literacy).

### 4.3.1.5   How the Development of Students' Software Literacy Impacts on the Teaching and Learning of Discipline-Specific Software in Formal Tertiary Setting

The development of software literacy occurred at various rates amongst students and was strongly shaped by lecturer teaching approaches, student expectations and disciplinary assumptions about the need to achieve professional levels of software competency. Many engineering students had little prior experience with SolidWorks (at high school), and this resulted in students learning at different rates, with a reliance on asking the lecturer as their preferred strategy for learning the software when at university (see Fig. 4.2).

Lecturer observations on student diversity touched on students' varying abilities and aptitudes including international students' prior software backgrounds:

> Some people go through very quickly, gifted students who can follow instructions if [they've used] drawing packages before will pick it up very quickly and they can finish in, I don't know, a tenth of the time … We accept that there is going to be a big gap between the best and the worst students, just like in maths or anything. You get this big distribution between the best and the worst and I don't think it's any different from any other subject.
>
> The difference between … let's say we have an overseas student come in and they've done some CAD but they may have done it in a very prescriptive manner. But when they come here

and are told, 'all right, just get on with that bit and we'll come and help you'. They're looking like frightened rabbits at the screen and not knowing what to do, so we spend more time with them to try and get them going. Whereas others, who haven't even used the software either, have just gone off, 'this is great,' and because they've got that passion for it, they're coming back the next week with it [their design] completed. So the gap between the one who's struggling and the ones who are at the top is just massive.

An aspect that appeared to facilitate students' learning of the SolidWorks software was students' prior engagement with artefacts or software that had a similar conceptual basis and so provided a pathway for them to engage more confidently with more advanced software such as SolidWorks. One student reflected on the value of playing with construction sets such as Lego:

Like people that mucked around with Lego and K'nex and that when they were younger, they're already on that wavelength, you know what I mean? And you'll find that helps heaps when you start going into 3D modelling, just because you sort of understand a little bit of how things go (Fourth-year engineering student).

Engineering students who had encountered 3D construction applications with similar sets of affordances to SolidWorks found it easier to pick up the skills to use SolidWorks. At least 19% of students reported using a range of software that had similar features to SolidWorks such as ProEngineer, AutoCAD, Star CCM+, Autodesk Inventor, TurboCAD, and Google SketchUp. They were able to comment on the similarities between these and SolidWorks in terms of their function (e.g., "They are designed for engineering related models/drawings"). Similar features between these different software were also noted such as "They have logical icons" and they have "sketch planes, extrudes, features, main interface".

Interestingly, just over half (54%) of these students were able to elaborate on the benefits of using these other similar software prior to their learning SolidWorks. These ranged from how the different software provided them with the "basic skills and familiarity" (e.g., "Helped to understand and get used to working in 3D on computers") in using CAD and "how software works in general" including more shared technical understanding of "reference geometry" and so forth. In focus groups students elaborated on the value of having a conceptual understanding of 3D modelling to be able to transfer skills across different CAD software:

Because SolidWorks is generally the first 3D computer programme that [students] learn and it's all about just getting the mindset of how you build something on the 3D programmes. So once you've learned the basics of what you want to do and how it's normally done you can find those features … it'll be called something else in a different programme but they'll be there. So in Pro Engineer it can all do the same stuff. In SolidWorks you could extrude a circle and then it's the same deal but it's called 'protrude' or something. All 3D modelling carries the same sort of [understanding] to be able to model stuff (Fourth-year engineering student).

Students who have had prior learning experience of other similar software therefore found the similar interface, logical and conceptual ideas inherent in those software helped them negotiate the learning of SolidWorks, and this, in turn, meant they could transfer many of the prior skills gained over when learning SolidWorks.

Students also proposed ways lecturers could approach the teaching of SolidWorks to help them demystify and become familiar with the software. Several ways were suggested. Firstly, as SolidWorks is a complicated application, students suggested a more in-depth grounding in conceptual frameworks in the learning of the software could facilitate their understanding and enable them to more effectively troubleshoot their application of specific affordances they encountered in their more informal learning. In the focus group, three students alluded to the need to be taught the overarching principles in terms of engineering design as well as CAD conventions to guide their SolidWorks use and enhance their understanding of the potential of the software:

> I think there are some things they can probably teach you more, like the use of planes and construction lines and stuff like that and then from there you can build on a lot of stuff, you know—if you've got a plane in the right place and orientate it to how you want it to be, it makes life a lot easier rather than trying to figure that stuff out (Fourth-year engineering student).

> It's the sort of software that you want to be taught right from the start how to do it properly and so otherwise you could spend so much time going running round in circles and building a big model, doing it totally wrong and then you spend a lot of time trying to fix it up. Whereas, you know, if you start from the start it's actually quite simple. You step back and think about what you're going to do and what's the best way to go about it (Fourth-year engineering student).

Other suggestions touched on the need to impart an awareness of the software's possibilities. Although some of this 'familiarisation' was achieved informally, students discussed the value of setting some expectations of what was possible to provide some benchmarks and motivations for their own (informal) learning. In the following comment, a second-year student linked this to working with a real-world case:

> I think what would be cool is if we had case studies or something; just some problems in class we could work through, the teacher could go through, like, "this is something that you may encounter while you're doing CAD, this is how we've gone about it, you could do it your way but this is the procedure we've used" (Second-year engineering student).

Other suggestions included being allowed more open-ended modelling assignments:

> Students have definitely got to muck around. They really struggle if they just went in, did the stuff and then just went home. I reckon a cool assignment would be to just take a household object and model it. Just give them [students] something and then get them thinking so they go home and think, 'Oh, can I model that? How would I do that?' Then they're thinking and then they would have to go and try to do it. Because I think a lot of people do that anyway but we need to get everybody doing it because I think you catch on real fast (Fourth-year engineering student).

To sum up, the student comments highlighted that, in order to cater for varying student abilities, experience and background (and different learning preferences), lecturers needed to use a combination of strategies (formal and informal) when teaching about and with software to facilitate the students' learning and development of software literacy. They thought that teaching the principles of engineering design

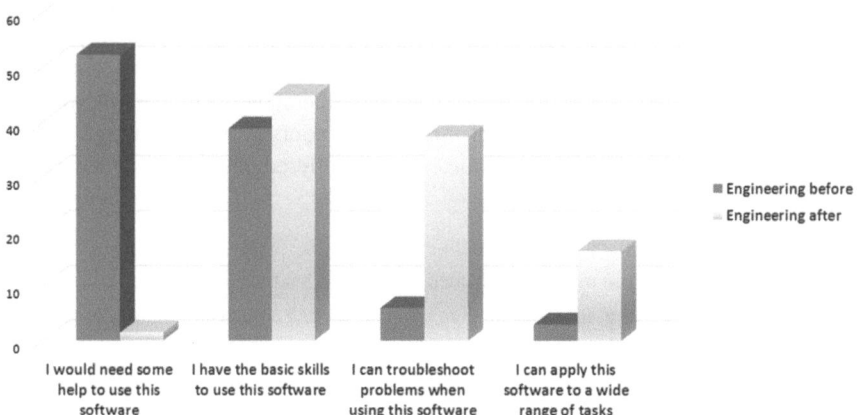

**Fig. 4.3**  Changes in engineering student assessment on their ability to use SolidWorks. [Adapted from Khoo et al. (2016), with permission from TLRI]

as well as CAD conventions and having to apply these understanding to solve real-world cases can enhance their understanding of the possibilities and potential of the SolidWorks software.

### 4.3.2  Student Perception of the Software Literacies that They Learnt as Part of Their Tertiary Coursework

Engineering students generally agreed that an understanding of CAD was necessary to comprehend and contribute to the engineering design process relevant to an organisation. Student evaluation of their ability to engage with discipline-specific software prior to and after completing their course indicated some gains in software literacy (see Fig. 4.3). Based on the categories of 'I would need help', 'I have the basic skills' (level 1 of our software literacy framework), 'I can troubleshoot problems' (level 2) and 'I can apply this software' (level 3), just over half of the students (52%) reported needing help to use SolidWorks initially before attending the course. This decreased to two percent at the end of the course. Also there was an increase from 39 to 45% of students who felt they now have the basic skills to use SolidWorks after learning about it in the course. Another 37% of students thought they were able to troubleshoot problems faced in using the software, an increase from the six percent who were able to do so at the beginning of the course. Gains in these two levels (basic skills and troubleshooting ability) correspond to the first two levels of our software literacy framework. However, by the end of the course, only 16% thought they could apply their skills to a wide range of tasks, an indication of a lack in achieving the third level of our software framework.

These results suggest that the formal coursework focused on software learning helped to develop students' software literacy so that nearly all students reported a shift to at least tier 1 (basic ability).

Apart from formal coursework, interestingly, another 27% of engineering students reported using SolidWorks outside of their formal coursework for a range of professional or recreational purposes. Some examples offered in the open-ended survey responses were:

> I have many sketches which I have a hard time imagining in 3D therefore I use it.
>
> [I use] SolidWorks to give me a more detailed version of what I have imagined.
>
> Designing a campus board for rock climbing training, playing around designing cars etc.
>
> Stress analysis of a conrod design for a model engine.
>
> Trying to design/modify something that has been in the market just for fun.
>
> General messing around. Attempting to design Iron Man suit.

Although very few students report achieved tier 3 of the software literacy framework from their coursework, having the basic skills to use and troubleshoot problems within SolidWorks was nevertheless an important part of preparation for the work place experience. Beyond the formal learning of SolidWorks in lectures and labs, the workplace experience offered engineering students authentic contexts to apply their knowledge of the software to solve real-world engineering design tasks. Two different students in the focus group explained the value of having at least an entry level CAD proficiency in the workplace:

> It is sort of expected to have some knowledge of CAD when you go into work placement. If you turn up with no background, it's a big disadvantage (Third-year engineering student).
>
> Because you'd always come across technical drawings so having an idea of how they're made can be a bit of a benefit especially if they're made wrong (Third-year engineering student).

For most students, their workplace required more specialised learning, faster and/or more complex levels of SolidWorks application to be more effective in addressing site-specific manufacturing/production processes. Hence different aspects of SolidWorks became more relevant than others for students' industry design purposes which extended their understanding of the software. This was exemplified when a student learnt a new application for SolidWorks as part of his workplace experience:

> I needed to do something and the boss pointed out another feature [in SolidWorks] that I had no idea, which was 'unpacking' or something. That opened my eyes to a whole different part, like there's an application that I had no idea existed and that I could do so much more with it (Third-year engineering student).

Another student commented on an example of using the 'virtual prototyping' feature in SolidWorks in his work placement to generate simulations of different design ideas and to allow his work team to discuss and decide on an idea:

> Yeah, so we'd use virtual prototyping if we needed to do a simulation to see how it [a design prototype] might behave under certain conditions. And then it was really good for when we

had multiple ideas on the table, they were all really good ideas but we needed the final sign-off by someone else so that's when it [virtual prototypes] came in (Third-year engineering student).

Furthermore, some work placements expected students to engage with similar but different CAD applications to SolidWorks requiring them to transfer their existing knowledge and proficiency to these contexts. A student gave the example of having to learn to use AutoCAD and other software such as Inventor as part of his workplace requirement. He found being exposed to the contrasting features of each software useful to his software literacy development:

AutoCAD's got more benefits because you can export your drawing to a Paint file and you can make it to a PDF and send it in an email to your boss. You can do all that from SolidWorks as well, it's just at university you're not taught any of that stuff in SolidWorks, there's limited knowledge of what you get taught and you only scratch the surface. My boss was saying using Inventor and AutoCAD, the benefits of AutoCAD is if you have a more complex model, if you want to make a last minute change to it, it's easier on AutoCAD (Third-year engineering student).

Finally, one student reflected on the overall value of learning to troubleshoot and of persistence when learning and using SolidWorks, be it in more advanced coursework or while on work placement:

From [first and second year] we pick up all the basic stuff and learn how to do it, but during that process we learn how to use the troubleshooting method and that's I think the most valuable thing that helped me later on … I'm confident with even something I don't know, I know how to find it, how to learn it from online resources then I can still make that happen [on SolidWorks]. I think that's the most valuable thing, that even later when I go to my fourth year and do some more complicated thing, I know where to go (Third-year engineering student).

In sum, students' reported learning of SolidWorks in their formal coursework, their own initiatives for using SolidWorks for personal and professional pursuits outside of university contexts, learning on the job (while on work placement) to use more specialised SolidWorks features, and transferring their learning of SolidWorks to the learning of other CAD software were strategies and opportunities that contributed to students' increasing sophistication and repertoire of software learning, and hence their software literacy development. Taken together, these views suggest that having a tier 1 and 2 level proficiency of SolidWorks was initially taken to be adequate for passing coursework and entry level into the workplace experience. However, there is an expectation that students will continue to enhance their software proficiency to complete specific engineering design tasks in their professional careers.

### 4.3.2.1  How Students Understand Software as an Influence on the Way They Encounter and Make Sense of Disciplinary Knowledge

As indicated in Fig. 4.3, a majority of engineering students reported shifting in their ability to use SolidWorks after learning and using it in their coursework but had difficulty identifying core disciplinary ideas embedded within software, or felt they were unable to critique the software they were using. Very few students in engineering discussed how SolidWorks shaped their disciplinary knowledge—a key part of software literacy.

However, the very few who are at tier 3 reported on the ways SolidWorks enabled them to visualise abstract disciplinary ideas, create and manipulate 3D objects, and communicate their design ideas efficiently to others as indicated in the following student quotes:

> You could say that you can make things in SolidWorks that you can't make in real life. So in SolidWorks you could [drill] a hole that was in a spiral and curve round but then you can't get a drill and drill that … that was a problem I came into when I was learning because I was just making models as they looked rather than how they could be made (Third-year engineering student).

> Probably the best thing is the integration, like you know how if you work in different parts or you've got different things you're working on and very complex things, it might be hard to put it all together on paper and see how it fits as a whole. [With the software] you can figure out how everything's going to work before you actually sort of build it (Fourth-year engineering student).

> And also like the supplier integration [feature]—the fact that you can build a part in Solid-Works and send it to someone who works with SolidWorks and they can send it for machine manufacturing, so there's a lot less chance of errors. And it's just fast to send things around … It's the whole package [from actual plans to the manufacturing process itself], rather than building a prototype of something, you can model it in 3D. It saves time and speeds up the process between design and manufacture (Fourth-year engineering student).

One student explicitly raised the need to have a clear plan for designing an artefact (i.e., an understanding of engineering disciplinary ideas about designs) in parallel with an understanding of SolidWorks' affordances in order to use the software productively:

> Just a clear plan of how it's [SolidWorks] going to work to do it the best way. Because it's only going to do what you tell it to do, and if you tell it to do something in a horrible way it's going to end up really messy, unless you've got a clear plan from the start. That's one key thing you've got to think, learn how the programme works and then make sure you're thinking along that same line (Fourth-year engineering student).

Generally, reasons for the lack of tier 3 software literacy proficiency amongst students was postulated to be due to the complexity of SolidWorks and the time needed to gain proficiency with using it. In focus group interviews, students considered SolidWorks to be a complicated, comprehensive and flexible piece of software. It was therefore not feasible to try and fully understand the breadth and depth of its hierarchies of affordances during their tertiary programme. Student quotes exemplified this complexity:

> Because there's so many tiny little individual parts about understanding SolidWorks that you get past a certain point and suddenly you don't know how to mirror a 3D part (for example) (Second-year engineering student).

> I don't think anyone can say that they've mastered SolidWorks. You can master it in your field but there's just so many different add-ons that you just have to try and get good at stuff (Fourth-year engineering student).

### 4.3.3 How and in What Ways Lecturers Model Attention to, and Use of Different Affordances in Discipline-Specific Software

When interviewed, engineering lecturers indicated they had a general understanding of the affordances and of best practice for SolidWorks use. For example, they pointed out the value of using CAD software in revolutionising the engineering design process and practice:

> If I went back, say, twenty years and I was drawing, designing something, when I've thought it through and I sketch out that high level design, and I said that's what we're making, then that's what we make. Well, we go into the detail to make it but I don't suddenly go round and change it, because of the effort to change that means everything else has to change, so we're locked into a design. It didn't mean we didn't design things; we still did then because those were the tools we had available. But with the modern software, you can have a whole machine like this and just change one bit, shrink it, you can add this, do this, and it's sort of revolutionised things.

They could articulate a rationale for their own practices and the relevance of these to their discipline content knowledge. However, the use of particular software teaching strategies was tempered by their assumptions about the level of software literacy they felt students needed to be a work-ready graduate. Different lecturers articulated the disciplinary assumption that the range of contexts and software applications that engineering graduates would need to engage with is diverse and sufficiently complex such that students would be learning various discipline-specific software throughout their careers. Lecturers highlighted the ways the teaching and learning of Solid-Works are guided by engineering design principles and industry and customer-focus demands.

> The [engineering] design process is very disparate, everyone comes up with different ideas, it's creative and uncontrollable. So we have to apply discipline to it and the discipline is through a method. So if we follow like these steps through it, it helps us control it ... the thinking process, creative process is just not controlled, it's not rigid. So unfortunately for us engineers everyone doesn't say, 'Oh well great, you had a lovely time thinking [that] up'—they want this at the end of the day, it's got to work. So the idea is to say 'Look, yes, you can follow this method that's internationally accepted as a way that if you go through your design like this, you can end up with something that works at the end of the day' and that was it, really, in a nutshell. And then obviously we're talking about how much time they should spend on the project because if they're graduates, are they on a salary, they have to justify their existence based on money, what they produce. So we sort of just let them know about that. We do that right the way through the programme ... we give them a lot of design freedom, but the actual process is quite planned. So it's creative but within particular parameters still.

The extent SolidWorks affords students becoming aware conceptually of the way in which real world engineering as a discipline works was also the result of lecturers using teaching approaches that related real-world practice with software teaching:

> Because the fourth years, they're designing the car now, they've learnt new skills, they're now [...] companies, they send the [...] bits over to the companies that are being made

directly from the design that they've drawn in SolidWorks, into the [laser cutters] and then they come back and they put it together, so there's a direct link [with manufacturing].

However, student development of software literacy up to tier 2 of our software literacy framework was generally viewed to be sufficient in preparing students to be lifelong learners of discipline-based software. Representative tutor/lecturer quotes were:

I do not believe that we're going to teach you [address to students] everything here. It's got nothing to do with that. You will be doing different projects, different software, and different things all through your career and you just have to have a mentality that you'll learn new things (Engineering lecturer).

The bottom line is we have [students] who are very good at SolidWorks and when they left [university] they found they weren't doing SolidWorks anymore because it's a very expensive software. Companies were just doing sketching by hand and then sub-contract that out to specialists in SolidWorks. So it's not every graduate engineer that has to know this software—[it's enough to] know it exists, and have the basics. Once they [students] go into industry they could be project managers or anything, they know it's there, they know its capability and at least some are really good at it and some are ok, and that's enough for [working in the] industry (Engineering tutor).

Someone who's not good at CAD might be still excellent in the design, and fifty per cent of the paper is on the design. So if you're really good at design you're still going to get a good mark because you'll pass the exam with lovely ideas. So it's quite nice, the technicians can be very good at that technical side of it and get good marks and the person who's not very good at CAD could still come up with reasonable CAD to get marks for that, but will do well on the creative design side of it. So there's options for both in the paper, there's opportunity for both to do well and if you're good at both then you can really ace it (Engineering lecturer).

These disciplinary expectations and assumptions underpinned lecturers' practice and were played out in the teaching approach to SolidWorks learning. Engineering lectures and labs therefore tended to point out the general affordances of the software before focusing on the specific tasks and functions that the affordances enabled to address specific engineering design issues. These included referring to and encouraging students to draw from informal learning strategies to supplement their formal lab learning:

Top three [student learning strategies] … Just by experience, just doing it. That one is number one. Number two would be they're Googling a lot of info, using the Internet; so when they can't find something they're just Googling it and going onto YouTube, whatever forum, any way they can to solve it. Number three … tutors. We're helping, so they come to the class and they can put their hand up and ask.

To sum up, engineering lecturers and tutors understood that they could not provide fully immersive training in SolidWorks at the tertiary level. They sought instead to ensure that students had appropriate learning strategies to empower them to understand how to begin to apply those strategies to different contexts and to explore the software on their own terms.

## 4.4  Summary

This case study of engineering studies students' software literacy development indicates that although students considered themselves early adopters of technology, knowledgeable in the affordances and constraints of their discipline-specific software—SolidWorks—they tend to draw from formal learning strategies initially when learning the software, supplementing these with more informal learning strategies. Students and lecturers articulated a wide range of learning and teaching strategies that were underpinned by disciplinary assumptions and industry expectations. Students, however, lacked a critical awareness of the extent software shapes (and reshaped) their learning of disciplinary knowledge. They raised the fact that the complexity of SolidWorks and the time needed to develop proficiency hindered their achieving level 3 software literacy proficiency. Students' diverse backgrounds and varied software skills and expectations meant that flexible teaching approaches were needed to accommodate their learning needs. Very few students achieved level 3 proficiency and demonstrated a critical reflexivity towards using SolidWorks to extend their engineering design abilities. Chapter 5 revisits these findings and discusses their implications in the wider field related to software teaching and learning.

## Appendices

### Appendix 4.1: Software Literacy Survey for Engineering Students

Dear students,

You are invited to take part in this survey about how students learn software, specifically, **SolidWorks,** as part of your coursework. We are interested in your experiences and opinions so that we can improve student learning experiences. Your participation is voluntary and will have no bearing on your course grade in any way. Your answers will be kept confidential. Your lecturer will not know the identity of students who participated in this survey. This survey is part of a two year research project funded by the Ministry of Education. The aggregated results from this survey will be reported to your faculty and in academic journals and conferences.

There are five sections in this survey. It should take you approximately 10–15 min to complete.

Note: Questions marked with an asterisk (*) are required. We would appreciate your answering all the questions. Please do not take this survey more than once.

By clicking the "Next" button below you give your informed consent to participate in this survey.
Thank you.

Elaine Khoo, Craig Hight, Rob Torrens, Bronwen Cowie
Research Team

**Section 1. Your background information**

1. **Which paper are you currently enrolled in (please select the paper in which you were invited to participate in this survey)?**
   Please choose **only one** of the following:

   ☐ Engineering second-year
   ☐ Other

2. **Please indicate your age.**\*
   Please choose **only one** of the following:

   ☐ Under 18
   ☐ 18–21
   ☐ 22–25
   ☐ 26–30
   ☐ 31–35
   ☐ 36–40
   ☐ 41–45
   ☐ 46–50
   ☐ Over 50

3. **What is your gender?**\*
   Please choose **only one** of the following:

   ☐ Female
   ☐ Male

4. **Are you a domestic or international student?**\*
   Please choose **only one** of the following:

   ☐ Domestic student
   ☐ International student

5. **What is your first language?**\*
   Please choose **only one** of the following:

   ☐ English
   ☐ Māori
   ☐ Other: _____

**Section 2. Your SolidWorks experience before attending this course**

We would like to know more about your views and experience with SolidWorks before attending this course.

6. **Which of the following best describes you?\***
   Please choose **only one** of the following:

   ☐ I love new technologies and am among the first to experiment with and use them
   ☐ I like new technologies and use them before most people I know
   ☐ I usually use new technologies when most people I know do
   ☐ I am usually one of the last people to use new technologies
   ☐ I am skeptical of new technologies and use them only when I have to

7. **Had you <u>heard</u> of SolidWorks <u>before</u> coming into the Engineering programme at uni?\***
   Please choose **only one** of the following:

   ☐ Yes
   ☐ No

8. **Please tell us how you first <u>heard</u> of SolidWorks:**
   Please choose **all** that apply:

   ☐ High school/secondary school
   ☐ Earlier university courses (e.g. year 1 course)
   ☐ Talking to lecturers in my programme
   ☐ Talking to more senior students
   ☐ Reading about it on the Internet
   ☐ Reading about it in books/journals/magazines
   ☐ Previous work experience
   ☐ Friends or family members
   ☐ Other: _____

9. **Had you <u>used</u> SolidWorks <u>before</u> coming into the Engineering programme at uni?\***
   Please choose **only one** of the following:

   ☐ Yes
   ☐ No

10. **Please tell us how you first learnt to <u>use</u> SolidWorks.**
    Please choose **all** that apply:

    ☐ High school/secondary school
    ☐ Earlier university courses (e.g. year 1 course)

☐ Working on projects with more senior students
☐ Working on personal projects
☐ Previous work experience
☐ Watched video tutorials on the Internet
☐ Other: _____

11. **Thinking back, how good were you in using SolidWorks before enrolling in this paper.\***
Please choose **only one** of the following:

☐ I would have needed some help to use this software
☐ I had the basic skills to use this software
☐ I could troubleshoot problems when using this software
☐ I could apply this software to a wide range of tasks

**Section 3. Your software learning from this course**

We are interested to know what helped you in your learning of SolidWorks as part of your coursework.

12. **Thinking back to when you were learning to use SolidWorks, please rank the strategies that were most useful to your learning of this software. Please rank from 1 to 9 to show the order of usefulness for your learning (from 1 = most helpful to 9 = least helpful).\***
Please number each box in order of preference from 1 to 9

☐ Ask the course lecturer/tutor
☐ Ask a friend/peer/other student
☐ Refer to the course lab notes
☐ Read a paper-based manual/step-by-step instruction booklet
☐ Go online/refer to the Internet for step-by-step instructions
☐ Go online/refer to the Internet for video tutorials (e.g. YouTube) to watch how to use it
☐ Watch someone using it in a face-to-face (physical) setting (not through videos)
☐ Discover through trial-and-error/practise
☐ Join an Internet forum (e.g. a discussion forum to ask other users for help)

12a. **Others (please share with us other strategies or resources you used to help you learn the software, if any):**
Please write your answer here:

_____

13. **Did you install SolidWorks on your own computer/laptop?\***
Please choose **only one** of the following:

☐ Yes
☐ No

14. **After learning about and using SolidWorks in this paper, how good would you rate yourself at using it?\***
    Please choose **only one** of the following:

    ☐ I would need some help to use this software
    ☐ I have the basic skills to use this software
    ☐ I can troubleshoot problems when using this software
    ☐ I can apply this software to a wide range of tasks

15. **Have you used any other software that you consider similar to Solid-Works?\***
    Please choose **only one** of the following:

    ☐ Yes (please tell us the name of the software)
    ☐ No

    Make a comment on your choice here: _____

16. **In what ways are the two software packages similar?**
    Please write your answer here:
    _____

17. **Did your having used the software you mentioned in Question 15 help in your learning of SolidWorks in this paper?\***
    Please choose **only one** of the following:

    ☐ Yes (please tell us how being able to use the software helped)
    ☐ No

    Make a comment on your choice here: _____

**Section 4. Your assessment of SolidWorks**

18. **In what way(s) does SolidWorks help you tackle an engineering problem?\***
    Please choose **all** that apply:

    ☐ Helps me visualise a solution that was in my mind in 3D format
    ☐ Helps me design and draw things before building them
    ☐ Allows me to draw an object to see what it looks like (or to share with others my drawing so they know what I mean)
    ☐ Allows me to easily modify my drawings
    ☐ Allows me to rotate/manipulate different angles or views of my drawings
    ☐ Helps to add details to my drawings so that they could be manufactured from
    ☐ Helps me explore effects of changes to measurements in my drawings

    Other: _____

19. **What does SolidWorks NOT let you do that you would like to be able to do?**
Please write your answer here:

_____

20. **Have you encountered any unexpected issues with SolidWorks?***
Please choose **only one** of the following:

☐ Yes (please describe the main issue you have encountered)
☐ No

Make a comment on your choice here: _____

21. **Have you used SolidWorks for your own personal interest/purposes out-side of coursework?***
Please choose **only one** of the following:

☐ Yes (please tell us how)
☐ No

Make a comment on your choice here: _____

**Section 5. We would appreciate your continued involvement**

**22. We would like to be able to follow up on how students learn about software packages in their university course. If you were interested in being part of this, we will be contacting you to ask a few short questions about your learning experiences with software within the next year. If you are willing to do this, please provide your student ID so that we can be in touch. Please note: Only the research team will see your survey responses. The team will not have access to any of your personal information or records kept in the university system.**
**My student ID is:** _____

23. **Would you be willing to take part in a group interview about your learning of SolidWorks?**
Please choose **only one** of the following:

☐ Yes (please provide your name and email and/or mobile phone so we can contact you)
☐ No

Make a comment on your choice here: _____

**Thank you for your time and help!**

# Appendix 4.2: Details of the Engineering Courses Investigated and Types of Data Collected

| Courses surveyed in the first year of the study | Data collected from the different participant groups |
|---|---|
| Engineering Design<br>(Year 2, Engineering students)<br>A second year course focused on engineering design, the design process, and group design projects for students to gain mastery of SolidWorks. Formal learning of SolidWorks was through lab-based learning followed by structured group project work to extend students' use of SolidWorks in engineering design | Data were collected from:<br>– 69 student surveys<br>– lecturer interview<br>– tutor focus group interview attended by four tutors including the lecturer<br>– lab observations of student learning of SolidWorks, and<br>– a student focus group interview attended by six students |
| Engineering work placement<br>(Year 2, Engineering student work placement in industry—on the job application of SolidWorks) | Data were collected from:<br>– four individual student interviews regarding student application and evaluation of SolidWorks in their work placement<br>– eight student assignments on their reflections of using SolidWorks in work placements, and<br>– a focus group student interview attended by seven students focused on their software experiences during work placement. Four other students who were unable to attend the focus group interview responded to the interview questions by email elaborating on their SolidWorks application experiences in the workplace. These email responses were coded alongside the individual and focus group interviews |
| Selected fourth-year students | Data was collected from a focus group interview with six elite final year students who have developed sophisticated engineering design and SolidWorks application skills |
| *Courses surveyed in the second year of the study* | *Data collected from the different participant groups* |
| Mechanical Engineering Design<br>(Year 3, Engineering students)<br>A third-year course focused on aspects of machine design where advanced engineering drawing and design techniques are further developed and applied through project work. Formal learning of SolidWorks involved advanced individual lab-based structured exercises and a real-world group project using SolidWorks | 47 enrolled students<br>Data were collected from:<br>– a lecturer individual interview<br>– lab observations of student project work involving SolidWorks, and<br>– a focus group interview with seven students |

# References

Arzi, H. J. (1988). From short-to long-term: Studying science education longitudinally. *Studies in Science Education, 15*(1), 17–53. doi:10.1080/03057268808559947.

Bell, P. (2004). On the theoretical breadth of design-based research in education. *Educational Psychologist, 39*(4), 243–253.

Johri, A., Teo, H. J., Lo, J., Dufour, M., & Schram, A. (2014). Millennial engineers: Digital media and information ecology of engineering students. *Computers in Human Behavior, 33,* 286–301. doi:10.1016/j.chb.2013.01.048.

Khoo, E., Hight, C., Torrens, R., & Cowie, B. (2016). *Copy, cut and paste: How does this shape what we know?* Final Report. Wellington: Teaching and Learning Research Initiative. Retrieved from http://www.tlri.org.nz/tlri-research/research-completed/post-school-sector/copy-cut-and-paste-how-does-shape-what-we-know.

Maykut, P., & Morehouse, R. (1994). *Beginning qualitative research: A philosophic and practical guide.* London, UK: Falmer.

Washington Accord. (2013). *Graduate attributes and professional competencies.* Retrieved from http://www.ieagreements.org.

Wertsch, J. (1998). *Mind as action.* New York, NY: Oxford University Press.

# Chapter 5
# Comparing the Cases: What Do They Tell Us About Software Literacy?

**Abstract** This chapter reports the comparative analysis of the two case studies on media studies software (see Chap. 3) and engineering software (see Chap. 4). Common themes emerged across the cases such as students' tendency to draw from informal learning strategies to supplement formal learning approaches, the diversity of student background and software abilities, and students' general assumption that a tier 2 software proficiency level (see Chap. 1) would be adequate entry into a professional pathway. However, the cases differed in terms of the nature of the nuanced learning goals and aspirations of each discipline which impacted on the way course curricular, teaching, learning and assessment strategies were structured. These findings have implications for teaching and learning where software plays a central role in understanding and accomplishing disciplinary ideas and practices in tertiary and workplace contexts.

## 5.1 Introduction

This chapter draws from the findings of two case studies reported in Chaps. 3 and 4 to investigate how discipline-specific software literacy develops in a formal learning environment and the extent this development fitted with our hypothesised 3 tier software literacy framework (see Chap. 1). The intention of the study was to unpack if and how students develop and use discipline-specific software, understand the influence of software on the way students make sense of disciplinary knowledge and whether their learning trajectories were consistent with the hypothesised tiers of software literacy. The chapter begins by highlighting similarities and differences between our case studies. It then offers recommendations and ideas for consideration in software teaching and learning practice and policy.

## 5.2 Comparing the Cases

The project described in Chaps. 3 and 4 of this book aimed at investigating the notion and development of tertiary student software literacy. We proposed a

© The Author(s) 2017  
E. Khoo et al., *Software Literacy*, SpringerBriefs in Education,  
https://doi.org/10.1007/978-981-10-7059-4_5

three-tier framework of development as a response to the ubiquitous but often neglected role that software plays across various sociocultural contexts. Two case studies were developed for engineering and media studies, based on collaboration with lecturers who were keen to examine the notion of software literacy. In both cases the focus was on the teaching and learning of commonplace discipline-specific software—Final Cut Pro and Adobe Creative Suite (media editing applications) in media studies, and SolidWorks (a computer-aided design, or CAD, software) in engineering. Both programmes are characterised by medium-sized enrolments of students with diverse backgrounds at entry level but differed in terms of the professional pathways for graduates. Both programmes used laboratory-based formal training in the learning of discipline-specific software, and provide resources for additional informal learning of software.

The learning and teaching strategies articulated by lecturers were underpinned by discipline-based assumptions and industry expectations. Within engineering, CAD has allowed engineers to engage in 3-dimensional modelling as a dynamic collaborative design process. While its use was assumed to enhance visualisation and communication and at the same time reduce unnecessary abstraction in the design process this was paired with the potential to circumscribe thinking and disincentivise the making of changes. Knowledge and proficiency of SolidWorks as commonplace CAD software is compulsory in our university learning context and would be expected of engineering graduates worldwide. New engineers are expected to develop non-technical skills such as communication, collaboration and entrepreneurship (see Washington Accord, 2013) and strategies that will allow them to continue learning outside formal contexts. Consequently, engineering lecturers focused on the need for students to have reached a level of proficiency (rather than to have mastered) CAD software when they graduate, which is equivalent to tier 2 in our framework. Lecturer comment indicated that this level was acceptable for graduation given the situated nature of the diversity of professional work contexts within which new engineers will be based. In addition, there is an expectation that new/emerging engineers will take up continual professional development activities to enhance their practice.[1] Within the focus of formal CAD learning some lecturers emphasised the importance of foundation subjects, such as maths and physics, while others emphasised design principles and industry and customer requirements. This latter focus positions the affordances of CAD closer to the centre of curriculum and better supported students to reach tier 3 of our framework. The few students we identified at this level indicated that CAD assisted them to visualise abstract ideas, create and manipulate 3D objects and communicate designs efficiently. It would seem these students were able to critically consider SolidWorks' affordances and apply this effectively to illustrate disciplinary

---

[1] As part of gaining chartered professional status and/or to become a full member of the Institution of Professional Engineers New Zealand (IPENZ) new/emerging engineers in New Zealand are required to submit a portfolio of work samples for assessment and undergo further testing to evidence they have gained sufficient experience. As a full member of IPENZ (commonly achieved 4–5 years beyond graduation), engineers still need to retain their chartered status by undergoing periodic reassessment to ensure they keep up-to-date with developments in their field and are adopting best-practices.

ideas about designs while engineering design and process lecturers offered students with parameters to guide their learning and experimentation with software.

As pointed out in Chap. 2 DNLE is now standard in media production and reframed ways in which this production takes place. It is located in an ecosystem of production tools which offer a wide palette of possibilities to media producers. Creative possibilities arise because of the ways various kinds of media can be layered within an overarching timeline. Additionally, DNLE has redefined the notion of editing allowing for multiple possibilities in the wider field of media studies through more direct user intervention, an integrated approach to combining sound and image, increased speed and efficiency in the use of digital workflows for editing audio-visual media.

In media studies (in the university which was the focus of this case study) the learning of DNLE was an elective. Media studies was depicted as having a more explicit focus on software literacy to complement a somewhat diffuse set of requirements including creativity, innovativeness and *accuracy*. Notions such as performative learning and critique of software tended to be made more explicit in media studies teaching. Tier 3 proficient media studies students (based on our framework) pointed out the ways their media editing and production software afforded freedom and versatility to create a wider range of aesthetic designs including working across different spaces, modalities and dimensions in new and interesting creative ways. Media studies lecturers sought to facilitate student developing appropriate learning strategies to enhance their understanding and application of these strategies to different contexts and to more independently explore the software's affordances and possibilities. The differences in expectations and assumptions played out in sometimes quite distinct ways for students. For example, students' engineering projects were more likely to focus on the design process with the aim of students developing a clear and coherent design *solution*, before implementing and testing this in a software environment. In comparison, media studies students were encouraged to explore creative possibilities by engaging with software, and to use a software more explicitly as a platform for experimentation and generating multiple versions of media content.

Irrespective of these differences, students in both disciplines considered themselves to be early adopters of technology and were generally comfortable in engaging with technology. They could identify and unpack the affordances and constraints of their discipline-specific software. In both programmes they could reach the stage of being able to troubleshoot problems they encountered in using the applications, and even suggest ways such software could be improved to support learning and use.

Students in general tended to draw from informal learning strategies to supplement more formal learning approaches, though for engineering students this approach was complicated by the fact that they were less likely to encounter CAD software before beginning their studies and had less access to these kinds of software off-campus. Commonly acknowledged informal strategies included asking more knowledgeable peers, using the internet to look up specific functionalities of a software either through YouTube or general search of specific forums dedicated to software learning, referring to a software's in-built tutorial in the case of SolidWorks, learning through trial-and-error, and so forth. These findings resonate with the wider scholarly findings examing student digital literacy (e.g. Alexander, Adams Becker, Cummins, & Hall Giesinger, 2017; Peeters et al., 2014).

However, our participating students in general lacked a critical awareness of the extent software shapes (and reshaped) their learning of disciplinary knowledge. Factors such as the complexity of discipline-specific software meant that considerable time was needed for students to learn and develop proficiency with their use. The few students who reached tier 3 software literacy proficiency were largely those who had, to some extent, already engaged with artefacts and software that shared some conceptual underpinnings with the applications they were required to learn in their coursework. They then had a basis for critique of their course software and were able to effectively tap into a software's affordances to extend their creative or engineering design abilities. Our findings align with the wider body of works from digital literacy calling for students developing more critical reflexivity when engaging with digital and software applications (e.g., Goodfellow & Lea, 2014) although these may not have an explicit focus of software as an actant in shaping disciplinary knowledge and action (Kitchin & Dodge, 2011) (also see Chap. 1).

## 5.3  Considerations and Recommendations for Policy, Practice and Further Research

Based on the findings from the study, the following recommendations are made for practice, policy and further research in terms of the potential of the three-tier software literacy framework, managing student diversity and disciplinary assumptions and nuanced nature of software teaching and learning contexts. These are elaborated next.

### 5.3.1  Support for the Three-Tier Software Literacy Framework

The findings from this small exploratory study support the existence of our hypothesised three-tier software literacy framework. However, student development and movement between the tiers was more fluid and flexible than we hypothesised. Student ability to achieve a higher level of software literacy does not necessarily preclude them from needing to revisit earlier levels in contexts where they encounter new but similar software. Developing a relatively sophisticated and critical understanding of an application or platform does not necessarily transfer directly to the learning of other forms of software. There are evidently a variety of factors in play here, only some of which we encountered in our research, such as engineering students reporting the need to re-learn certain functions of SolidWorks to complete a new task more efficiently while on work placement. In practice, the tiers are not necessarily distinct but rather the boundaries between these are permeable.

Although the shift from basic skills in using affordances (tier 1) to the ability to independently trouble-shoot applications (tier 2) involved reaching a clear threshold for most students, the transition to a more critical understanding of software (tier 3) was less easily demonstrated. As noted above, some students could understand and critique the conceptual frameworks underlying the applications they encountered in formal training, but did not always extend these skills to other software they encountered. As with other forms of digital literacy, much depends on a variety of factors informing and shaping each students' learning. This suggests lecturers should not assume student competency across contexts (e.g., informal to formal, from campus to workplace settings) or across similar but different media editing software such as Adobe Photoshop and Adobe After Effects, for example. Nonetheless, our framework has value as a conceptual tool for practitioners in terms of understanding the role of troubleshooting as an important development stage in learning with and through software. Understanding how to teach themselves the more complex possibilities afforded by an application—where and how to tap into resources such as built-in tutorials, YouTube videos, peers and tutors—is a valuable skill and, we argue, represents a key threshold for students to reach.

We see value in students' gaining tier 3 capability, as the ability to critique software is fundamental to understanding that software code is never 'neutral' (Fuller, 2008; Manovich, 2008); as outlined in Chap. 1 it is a form of writing which informs and shapes possibilities for action. Students who are able to transfer a critical understanding of software affordances have the sophisticated understanding needed for considering how software enable some kinds of knowledge and actions while also potentially constraining other forms of knowledge and actions. These students are able to use new software and familiar software in new contexts and situations because they understand the conceptual framework that underpins a software. This said, the affordances and constraints of discipline-based software in shaping disciplinary knowledge needs further consideration by students and by lecturers with lecturers needing to be aware of the implications of their choice of software and modelling of a software application.

Even when evidence of students achieving tier 3 exists, it cannot be assumed that they have mastered all facets. Compared with tiers 1 and 2, tier 3 is more complex and therefore difficult to achieve. Its multifaceted demands play out in multiple ways to the extent that we consider the skills and understandings required for tier 3 literacy vary according to the demands of the particular discipline and the task at hand. For example, in media studies critical thinking tends to be seen as a core aspect of creative disciplinary knowledge. Students' development of reflexivity is as important as the development of the capacity to produce a creative product. Tier 3 ability therefore is essential for media studies graduates to be competitive in their profession. Being able to judge the creative capacities of competing software has implications for the nature of the practice which is developed, the form and eventual conceptual complexity of media products themselves, as well as for more practical considerations such as budget. In contrast, in engineering the disciplinary emphasis is on the quality of an engineering design—which typically means adhering to objective and external measures of successful design, not least of which includes the satisfaction of an initial client's brief. For engineering graduates, design efficiency and effectiveness is prioritised over the kinds of creative reflexivity observed in media studies (i.e., the function of a product is more important than the form). Nonetheless, we consider tier 3 would be needed by experienced engineers charged with selecting and/or recommending different software as fit for a particular industry's purpose/task.

Our findings indicate there is value in each discipline examining how discipline-specific software teaching and learning is positioned in relation to graduate profiles. Software teaching and learning environments where students encounter a range of competing software tools would be needed to raise awareness of the affordances of different discipline-specific software. This, in addition to learning a discipline-specific software in-depth, can be beneficial to providing students with the breadth and depth of literacies required to engage successfully with software in their discipline.

### 5.3.2  No One-Size Fits All Approach to Software Learning

Multiple learning pathways exist for exploring the affordances of any particular software, both formal and informal. Students often prefer informal strategies as a supplement to, and at times above formal strategies for learning discipline-based software. Learning software can be a complex and challenging undertaking, and is dependent upon the quality of cases and projects which are the immediate focus of the learning. Lecturers could usefully be informed by, and take advantage of students' informal repertoire of learning strategies and networks, including their accessing web-based resources and discussing with peers. These are all integral to developing troubleshooting strategies and acquiring an understanding of the creative possibilities of applications and platforms. Lecturers drawing from students' already established informal learning strategies should recognise the relevance of the social and cultural context in shaping effective technology and software engagement. Our findings indicate international students from cross-cultural backgrounds are acquainted with more prescriptive methods of software learning, for example, would need more assistance to become confident users of discipline-specific software as compared to students who have gained software skills through town initiatives in experimentation with a range of similar software.

Our findings add to debates over assumptions that young people/current students are a digitally literate generation. Our participants perceived themselves to be competent and confident early adopters of technologies. Many graduating students were able to recognise discipline-specific software affordances and acknowledge these to be central in their engagement with disciplinary knowledge. However, very few were able to critique how the software might *shape* their disciplinary knowledge. In focus groups centred on discipline-based software learning, very few students demonstrated critical thinking about the nature and role of the software they were using and most were not able to describe or discuss applications for software beyond those used in their learning. That is, there was very little evidence of tier 3 software literacy. Lecturers need to explicitly teach and model software critique if they wish to foster this capacity and/or make this possibility known to students. That is, lecturers need to encourage students to engage with and think about the content and presentation of their designs created with software and how this influences students' interpretations and engagement with disciplinary knowledge.

As with all learning, the diversity of student cohorts and the range of understandings and the variation in familiarity, skills and experience students bring with them to the formal software learning context constitute a further challenge for teaching of and through software. Some students may already have a critical orientation towards software, or have acquired an understanding of the conceptual framework underpinning an application. Our findings indicate an advantage in terms of more advanced software learning for those who have prior experience with other software with a similar conceptual framework. In response to this diversity, lecturers could usefully direct time and attention to formatively assessing students' initial software literacy and adapting teaching activities in light of this. This said, our findings indicate there is no single best (one-size fits all) approach to teaching discipline-specific software. Lecturers adopting a range of teaching approaches (formal and informal) and being flexible to address diverse learning needs represents a crucial part of supporting student learning. We recognise tensions in terms of time and depth versus breadth of ideas that each lecturer will need to address, and hence reiterate our earlier point for a re-examination of where and how software-based literacies are positioned within each discipline.

### 5.3.3 Situated Nature of Software Teaching and Learning

We advise caution in the interpretation of our findings because of the need to consider the situated nature of our investigation of discipline-specific software learning. The participants in the study represent a convenience sample of lecturers and students from one educational institutional setting. They were from two distinctly different disciplinary contexts, with distinct and different disciplinary foci and expectations of software teaching and learning. Lecturers were also careful to point out the study only focused on some courses within a programme and all universities have different interpretations of how software teaching and learning can be enacted. Although the findings cannot be generalised, we hope that by providing *rich thick descriptions* of the context and action (Lincoln & Guba, 1985) (also see Sect. 3.2.4 for the ways the study has sought to add rigour in its examination of the cases), readers will be able to draw insights for their own uses from the study.

## 5.4 Summary

This chapter synthesised and compared the key findings across two case studies into students' software literacy development in formal learning contexts. The key findings suggest lecturers would be wise to introduce and help students to develop critical awareness of how software can inform and shape their understanding of disciplinary knowledge and practice. The chapter further offered recommendations for policy, practice and further research. Limitations of the study were also reported to highlight the nuanced and situated nature of discipline-specific software teaching and learning.

# References

Alexander, B., Adams Becker, S., Cummins, M., & Hall Giesinger, C. (2017). *Digital literacy in higher education, Part II: An NMC horizon project strategic brief* (Vol. 3.4, August 2017). Austin, TX: The New Media Consortium.

Fuller, M. (2008). *Software studies: A lexicon.* Cambridge, MA: MIT Press.

Goodfellow, R., & Lea, M. R. (2014). *Literacy in the Digital University: Critical perspectives on learning, scholarship and technology.* New York: Routledge.

Kitchin, R., & Dodge, M. (2011). *Code/space: Software and everyday life.* Cambridge, MA: MIT Press.

Lincoln, Y. S., & Guba, E. (1985). *Naturalistic inquiry.* Beverly Hills, CA: Sage.

Manovich, L. (2008). *Software takes command* (online draft). Retrieved from http://softwarestudies.com/softbook/manovich_softbook_11_20_2008.pdf.

Peeters, J., Backer, F. D., Buffel, T., Kindekens, A., Struyven, K., Zhu, C., & Lombaerts, K. (2014). Adult learners' informal learning experiences in formal education setting. *Journal of Adult Development, 21*(3), 181–192. doi:10.1007/s10804-014-9190-1.

Washington Accord. (2013). *Graduate attributes and professional competencies.* Retrieved from http://www.ieagreements.org.

# Chapter 6
# Software Literacy: Education and Beyond

**Abstract** Software literacy is an essential part of learning and living in the 21st century; something which, we argue, transcends the use of any particular tool and any particular educational, social and cultural context. Software literacy is an increasingly central part of the palette of understandings and skills that comprise the broadening umbrella of digital literacy. It is therefore essential that citizens have a critical understanding of software to make more informed choices about their use, can transfer this critical understanding to software they have yet to encounter, and understand that all software has nuanced affordances and limitations. In tertiary settings this is needed to ensure equitable and critical learning with and through software. This chapter summarises our key insights from our own research into these issues and offers recommendations for future research in the field.

## 6.1 Introduction

The need for a critical understanding of the role and significance of coding needs to be acknowledged as a core part of the palette of understandings and skills that comprise the broadening umbrella of digital literacy. The need to understand software transcends the use of any particular tool and any particular educational, social and cultural context. As a cultural artefact, coding plays a role in reproducing, reinforcing and augmenting existing cultural practices, as well as generating new practices. The infrastructures in which code is embedded constitute an increasingly pervasive presence in everyday society, one that mediates, supplements, augments, monitors and facilitates individual and collective activity. Software such as Google, the iOS software in iPhones and iPads, and Microsoft Office software packages are just everyday examples reflecting the extent to which software has become embedded in everyday personal and professional pursuits. It is therefore both desirable and advantageous that citizens have a critical understanding of software to make more informed choices about their use, can transfer this critical understanding to software they have yet to encounter, and understand that all software has nuanced affordances and limitations. Our own small contribution to empirical research in this field demonstrates some of the implications for education providers, especially universities, in their role in

© The Author(s) 2017

E. Khoo et al., *Software Literacy*, SpringerBriefs in Education,
https://doi.org/10.1007/978-981-10-7059-4_6

fostering critical thinking and serving as critic and conscience of society. It is crucial to ensure all students and lecturers are supported in teaching and learning processes whether these are mediated through and/or focused on software.

## 6.2  Software as a Digital Literacy

First let us return to, and extend some key principles established in Chap. 1. These were drawn largely from recent thinking from Software Studies scholars, Manovich and Kitchin in the main, who argue and illustrate that we are living in a software culture which is fundamentally (re)shaping all areas of modern life. The principles are:

- The conceptual framework inherent to software is not neutral, and understanding this is central to critical literacies across a range of domains. This principle suggests the need to ask questions such as: Do we understand specific pieces of software, how they operate, what they do and do not offer us, and are we conscious of how specific pieces of software inform our imagination of what is possible?
- Our *performance* of software is intimately connected to our understanding of its underlying conceptual framework prompting the need to consider: Do we adapt ourselves to it, do we push it to do things it was not designed for but is capable of doing (do we *hack* it?), do we put it aside and look for a better solution, can we combine it with other software to accomplish the current task and/or to imagine new things?
- Software translates practice into something new—perhaps in scale, perhaps in terms of who can and wants to participate and how they can participate, and perhaps in going beyond the individual to invite others to participate as a collective with tools like wikis/Wikipedia, and open source codes that allow public/masses to contribute their thoughts to improve on ideas, tools and practices. Software is intimately associated with (new) forms of automation (see below) that allow us to amplify, extend and effectively create new kinds of practices.
- Software evolves, and not in necessarily linear or carefully designed ways—much of software is combinatorial, where something is coded (an affordance, an interface, a platform, or an interoperability between previously distinct systems) and this becomes a building block that extends an existing capability or can be combined with other building blocks to possibilities not always anticipated by the original coders.

Beyond these general principles, in our research and this book we have focused on software and software literacy in the context of engineering and media studies education. The term *literacy* directs attention to how individuals make use of software and for what purposes, in what contexts and with what anticipated and unexpected outcomes.

Our three tier definition of software literacy includes understanding the technical as well as the conceptual and sociocultural aspects of software use. It includes operational aspects, the capacity to problem solve when using software to complete a task, and a critical awareness of affordances and the implications of these for the conceptual framing, practical approaches and values a software supports. Whereas the definitions of other kinds of literacy focus on the general societal, social, communicative and creative aspects of literacy, our initial focus within software literacy is deliberately on applications which have implications for our creative agency, and play out in concrete ways within disciplinary and professional practices. As discussed in Chap. 1, these have consequences for our agency as creative producers. Our conceptualisation of software literacy focuses in on user-software interactivity, more specifically what the affordances software can enable and constrain and the implications of this for notions of agency within human-machine assemblages embedded within software culture. In line with current thinking in Software Studies, our focus is on how users and these assemblages evolve together in ways that both proscribe and expand how tasks can be and are conceptualised and accomplished. At its most sophisticated our conception of software literacy raises the need for users to question the implications of their use of a particular software where these implications could be social, ethical, cultural, legal and/or practical and pragmatic.

## 6.3 Software, Software Literacy and Education

Outside the more general principles developed from Software Studies detailed above, Kitchin (2015) has set out the variety of ways in which software has become ubiquitous within educational contexts, typically in unexamined ways. He lists these as teaching materials that are created using software programmes, teaching that is co-delivered through digital media, and assessments that are conducted using software packages. Kitchin's focus is very much on the role software plays in affording different approaches to teaching, learning and assessment from a teacher's point of view. We are also interested in the implications of software use for curriculum knowledge and the way these are demonstrated which in turn has implications for what it means to be a learner and knower. Buckingham (2007, p. viii) argues that, "we need to be teaching about technologies, not just with or through them". We agree with him when he points out: "form, media or platform do not speak". The way that content is mediated, framed and shaped by the platform and software selection needs to be made explicit and critiqued as part of the educational process.

Across Chaps. 3 and 4 we have set out case studies on how the professional software students use shapes, and frames what is thinkable in a discipline. The variations in students' location on our three tier software literacy framework reinforces that what is critical is the person-software assemblage. Here what is at stake is both curriculum and what students come to understand as a valid and valued professional identity.

Our research begs the question of the need for coding skills to be included in formal curricula. Internationally the answer *yes* is gaining traction (e.g., Williamson, 2015). Vee (2013, 2017), for example, argues there is a parallel with mass ability to read and write and predicts equally pervasive changes to society when most computer users have the knowledge and power to create and modify software. Computational thinking, a similar idea, refers to the style of thinking used when programming a computer or developing an algorithm. It includes formulating, representing and analysing problems in terms of their component parts. At the core of debates over the need for coding to be taught as a key form of writing is an unease with a future where inequalities may increasingly centre not just on *access* to digital technologies but the ability to operate with meaningful agency in coded environments. This does not just mean a division between those who can code (i.e. write their own software) and those who cannot. At a more fundamental level, it involves an effort to prevent students from being *captured* by software; from being unable to imagine possibilities outside of the tools they use on an everyday basis. As with much in this field, we have more rhetoric than actual empirical evidence to inform educators on concrete strategies to employ in the classroom. This then is a potential area for development in our software literacy framework.

As soon as we turn our attention to researching such matters we need to remind ourselves that just as digital technologies are now entwined with multiple aspects of educational practice they are also entwined with educational research practices. This entanglement ranges from the conceptualisation of a project to knowledge dissemination efforts. Digital devices, such as voice recorders, digital cameras, and/or iPads, are used during fieldwork. Dropbox and Google Docs are used for collaboration and data sharing. Software programs such as NVivo and SPSS assist with data analysis. Social media platforms, such as Facebook and Twitter, are both sources of data and channels for disseminating research findings. Beer (2012) points out that the operations these digital devices and interfaces perform, at our behest are powered by algorithms and codes. And that by "streamlining, making efficient, predicting, making decisions for us, doing work on our behalf, [they are] taking some of the agency from researchers and the research process and making it their own" (para 2). At the same time these software algorithms and codes are usually black-boxed, meaning that the calculations, decisions and processes being performed are essentially hidden from our critical analysis (Roberts, Hine, Morey, Snee, & Watson, 2013). Roberts and colleagues argue this leads to "insufficient attention to methodology" (p. 6). They suggest, and we agree, that the encoding of research practices via digital tools needs to be brought out of the background to consider how this is shaping—enabling and constraining—how educational research is performed.

Lynch (2015), in *The Hidden Role of Software in Educational Research*, adopts Software Studies as a paradigm to critique the long-standing tendency to adopt new technologies without considering the deeper implications of their use. Lynch develops and uses an interesting conception of *software space*, drawing in particular from Kitchin and Dodge's notion of *code/space* (Kitchin & Dodge, 2011). Lynch outlines how:

> I conceptualize *software space* to refer to a complex computational assemblage that includes devices, networking infrastructures, interfaces, code and information systems. Software space is socially situated between 1) political and economic spaces, which includes actors that create policy, produce software and promote its adoption as well as 2) educational and

administrative spaces, which include students, educators, and school leaders. As my evocation of spatiality is intended to suggest, the interplay between the different spaces is fluid and, often, inhumanly fast (Lynch, 2015, p. 25).

Lynch outlines a model for critical Software Studies, which he suggests as a means for investigating the "hidden ways issues of ideology, power, and inquiry are encoded in educational software spaces" (Lynch, 2015, p. 50), or what he calls (drawing from Manovich) the *softwarization of [U.S.] education*. In practice, this entails looking at software space at a variety of levels; devices, network infrastructure, interfaces, code, and information systems to consider the hidden political, economic and epistemological ways it impacts on practice.

The broader lesson here is important; while we have specifically studied examples of software as applications, as tools that wait to be interacted with, to be performed toward particular ends, we need to consider future developments of software (not least those which cede further agency to software-based infrastructures within education and educational research).

## 6.4  Software Literacies in a Coded Future

Software is embedded within many, even most, of the activities fundamental to our lives—work, leisure, social interaction, health and wider well-being. Here we discuss just a few of these to provide a context for thinking more broadly about the need for all of us as citizens to be software literate at the third tier of our literacy framework, irrespective of whether we have the skills and knowledge to perform software at tiers 1 and 2 (or indeed, any proficiency in coding).

A fundamental facet of software culture, one which facilitates the transformative potential of coded practices, is automation. Automation enables aspects of a practice to be translated into algorithmic form, and hence scaled up to whatever size is desired (limited only by available processing power). By combining different automated processes, sequentially or in parallel, software culture can start to exhibit practices that take on their own distinctive quality, and in ways that eventually become *naturalised* for software users (MacKenzie, 2006, p. 44). Semi-automation in Word processing, for example, includes deleting mistakes, cutting and pasting, counting the number of words, automated line numbering and basic spell-checking At a more everyday level, there are many kinds of operations which we only belatedly realise we cannot do without: the automatic focus on a smartphone camera, customisable notifications for all sorts of everyday tasks, automated mapping operations, and various forms of file management. Such automated affordances are central to the seductive power of software environments. They are all slowly allowing us to operate in a software bubble that is not always noticeable unless it fails to function. Our reliance on software-based automated functions is a key way in which we give some of our own agency to code.

At the platform level, Facebook and other social media are examples where users are engaging with largely automated systems, creating quite distinctive, *coded* communication practices. van Dijck's (2013) research into global social media platforms provides a valuable political economy of key parts of contemporary software culture, demonstrating that at this level code operates to provide a broader infrastructure of engagement, a set of *givens* for how users can engage, participate and interact. For example, the Facebook platform as a "friendship assemblage" (Bucher, 2013, p. 490) encourages particular kinds of behaviour, such as gathering large numbers of friends, and Facebook's *like* button shapes how people interact. Looking to the future Web 3.0, which Spivack (2007) defines as connective intelligence, instead of multiple searches, you might type a complex sentence or two in your Web 3.0 browser and various forms of personalised search infrastructures will do the rest. These capacities come with the potential to change how we conceptualise the shape of knowledge and what it means to know. At the same time it comes with the need to consider the algorithms that are informing what and how information/knowledge is marshalled and presented to us as comprehensive whereas the process may amplify certain information narratives whilst silencing others. In such scenarios, the ultimate consequence is arguably the limiting of opportunities for a person to encounter conflicting views (MacKenzie & Martin, 2016).

The way we have traditionally thought about the internet has been in terms of pages, but we are about to see this changing to the concept of *streams*. In essence, the change represents a move from a notion of information retrieval, where a user would attend to a particular machine to extract data as and when it was required, to an ecology of data streams that form an intensive information-rich computational environment. This notion of living within streams of data is predicated on the use of technical devices that allow us to "manage and rely on the streaming feeds" (Berry, 2011, p. 143). In this ecosystem, we are in the role of managers of complex streams of information generated through our engagement with largely automated practices.

Artificial Intelligence (AI) is the end-point of some of this thinking, in terms of an imagined sentient intelligence which is independent of user input, but there are a series of milestones along this trajectory that we are already living with—whether it is the massively complex field of algorithms which constitute the Google search engine, or YouTube recommendation system, or Apple's Siri digital assistant, or (less obviously) the manner in which the global share market system is deeply embedded within software practices—the economy, to a large extent, is *coded*. AI-centred search engines, linked to the Internet of Things, are offering a variety of services, such as real-time language translation, but also anticipating our requests and actions through predictive analysis. Military uses for everything from targeting systems, to AI-centred wearables to enhance the stamina and capability of individual soldiers, through to the development of gaming forms as training modules (e.g. building from early prototypes of software-based training environments, such as the America's Army first-person shooter (FPS) game: https://www.americasarmy.com/) are part of our current and prospective landscape.

At the same time the potential for virtual reality (VR) and augmented reality (AR) applications are only just being conceptualised, but already attract enormous investment from a variety of technology-centred sectors. The extensions of architecture and design outlined in Chap. 2 indicate an increasing movement into BIM (building information modelling), which, among other possibilities, allows architects to use VR platforms to walk through their simulated models. While VR provides a seamless engagement with a computer-generated environment, AR (and offshoots such as aural augmented reality (AAR)) aim to add information layers to our landscape, using interfaces to networked content through wearable devices such as goggles, headphones and more intuitive devices. The vision here (pun intended) is a redefinition of everyday space as globally networked, by default.

There are any number of Big Data-driven software developments, for example within political practices, such as identifying voting patterns, collating voter lists and their political tendencies, through to the influence of bots designed to disrupt the communication strategies of opposing campaigns and exploit the power of Twitter and other social media platforms outside of mainstream journalist professional practices. Further developments in biometric monitoring, knitting together personalised wearable devices and coded infrastructures emerging within smart-homes as part of the Internet of Things, offer the potential for self-surveillance of all sorts of measures and data on health and illness, ultimately to be networked to automated diagnostic tools. Any number of other health-related applications of Big Data and mobile technologies are currently being developed, not least the more efficient and real-time gathering and analysis of data to identify the emergence of new viruses, and to assist in rapid responses to restrict their spread. In this environment of intensifying software developments, and crucially the emergence of automated, real-time, customised and naturalised coded practices, software literacy becomes core to our civil roles and responsibilities within society.

## 6.5 Concluding Comments: Towards Digital Citizenship

Digital technologies are introducing new possibilities for what it means to be a citizen. People are less tied to the demands and restrictions of a specific place. To be a digital citizen is to be more globally aware, more critical, more willing to challenge the immediate (Bennett, Wells, & Rank, 2009; Coleman, 2006; Hermes, 2006). Advanced citizenship is highly dependent on agency where the individual needs to be part of and contribute to a community/multiple communities. In order to understand the role of individuals in civic life in the shadow of code and all its manifestations, it is imperative to gain a firmer understanding of what activities it engenders, the ways it changes the perceptions of public issues, and shapes ways of working and generating knowledge.

It is difficult to offer predictions about the kind of digital/software environment we will experience in the next five to ten years without lapsing into science fiction. The perfect storm of artificial intelligence and machine learning, for example, is often predicted to feed into all forms of the economy, and suddenly render entire professions defunct. There are a host of technologies becoming mature, and feeding off each other to create new possibilities—collectively these offer the potential to deepen and make more subtle, complex forms of human-software symbiosis. The question of who controls the software will become an increasingly pressing issue—both immediately in terms of the effective control of existing operating systems, and software-based ecosystems represented by corporate entities such as Google, Apple, Facebook and a comparatively small number of players dominating software culture. This is intimately connected with larger issues to do with the nature of human agency and accommodation with software-based systems and infrastructures which are *assisting* with human decision-making. And this in turn feeds into current concerns over the kinds of skills and competencies that young students should be focusing their energies and talents into. At the heart of such concerns is an awareness of the power relations associated with software culture—an aspect which we might put into *critical software literacy*. These feed into a host of ethical concerns over how to identify responsibility for actions within human-machine assemblages (e.g., the debates over insurance and other issues related to automated driving which is holding up their release to the market).

Software is only going to become more deeply embedded in the fabric of everyday life with many people predicting increasing deeper and more powerful integrations of human and machine assemblages. Key questions arising from this are: How are people making decisions? How empowered are they or not through their exchanges of software? Ultimately, we return to the broader challenge we posed at the beginning of this volume: why and how does software matter?

In the immediate context of our research, such questions should serve prompts for education providers, especially universities, to re-examine their role in fostering critical thinking and serving as critic and conscience of society. It is crucial to ensure all students and lecturers are supported in teaching and learning processes whether these are mediated through and/or focused on software. As an increasing range of workplaces make use of software and educators move to exploit the potential of e-learning platforms, and make use of social media and cloud-based and mobile applications, our research highlights the need for further detailed empirical investigation of software literacy. In tertiary settings, this is needed to ensure equitable and critical learning with and through software.

# References

Beer, D. (2012, October). *Algorithms in the academy: The sorting of academic practice [Web log message]*. Retrieved from http://thinkingculture.wordpress.com/2012/10/20/algorithms-in-the-academy-the-sorting-of-academic-practice/.

Bennett, W. L., Wells, C., & Rank, A. (2009). Young citizens and civic learning: Two paradigms of citizenship in the digital age. *Citizenship Studies, 13*(2), 105–120.

Berry, D. M. (2011). *The philosophy of software: Code and mediation in the digital age*. Houndmills, UK: Palgrave Macmillan.

Bucher, T. (2013). The friendship assemblage: Investigating programmed sociality on Facebook. *Television & New Media, 14*(6), 479–493.

Buckingham, D. (2007). *Beyond technology: Children's learning in the age of digital culture.* Cambridge, UK: Polity.

Coleman, S. (2006). Digital voices and analogue citizenship: Bridging the gap between young people and the democratic process. *Public Policy Research, 13*(4), 257–261.

Department of the Army (n.d). *America's Army* [computer software]. Available at https://www.americasarmy.com/.

Hermes, J. (2006). Citizenship in the age of the internet. *European Journal of Communication, 21*(3), 295–309.

Kitchin, R. (2015). Foreword: Education in code/space. In B. Williamson (Ed.), *Coding/learning: Software and digital data in education* (pp. 1–3). Stirling, UK: University of Stirling.

Kitchin, R., & Dodge, M. (2011). *Code/space: Software and everyday life.* Cambridge, MA: MIT Press.

Lynch, T. L. (2015). *The hidden role of software in educational research: Policy to practice.* New York, NY: Routledge.

Mackenzie, A. (2006). *Cutting code: Software and sociality.* New York, NY: Peter Lang Publishing Inc.

MacKenzie, A., & Martin, L. (2016). *Developing digital scholarship: Emerging practices in academic libraries.* London, UK: Facet Publishing.

Roberts, S., Hine, C., Morey, Y., Snee, H., & Watson, H. (2013). *Digital methods as mainstream methodology: Building capacity in the research community to address the challenges and opportunities presented by digitally inspired methods.* Retrieved from National Centre for Research Methods website: http://eprints.ncrm.ac.uk/3156/.

Spivack, N. (2007). *The semantic web, collective intelligence and hyperdata.* Retrieved from http://novaspivack.typepad.com/nova_spivacks_weblog/2007/09/hyperdata.html.

van Dijck, J. (2013). *The culture of connectivity: A critical history of social media.* Oxford, UK: Oxford University Press.

Vee, A. (2013). Understanding computer programming as a literacy. *Literacy in Composition Studies, 1*(2), 42–64.

Vee, A. (2017). *Coding literacy: How computer programming is changing writing.* Cambridge, MA: MIT Press.

Williamson, B. (2015). A hidden computing curriculum. In B. Williamson (Ed.), *Coding/learning: Software and digital data in education* (pp. 92–98). Stirling, UK: University of Stirling.

# Glossary of Key Terms

**Affordance** A person's perceived opportunity to utilise a particular tool for action; for example, a doorknob is for turning. Within software, these are typically organised through user interfaces.

**Coding (or programming)** A form of writing which inscribes types of actions to be performed using a computer.

**Digital literacy** An umbrella term for a number of competencies, skills and understandings associated with digital infrastructural technologies (we argue software literacy falls within this same umbrella).

**Pedagogy** Study and practice of teaching.

**Software** Machine readable instructions which direct a computer's processor to perform specific operations. For everyday software applications, these instructions are put into operation as computer users run an application.

**Software application** A computer program designed to perform a set of tasks or functions. Common everyday examples range from an *app* on a smartphone, to desktop computer programs such as a web browser.

**Software literacy** The repertoires of skills and understandings needed for students to be critical and creative users of software applications and systems in a software saturated culture.

**Software platform** A coherent programming environment or system which supports applications. Examples include operating systems such as iOS or Android, or social media platforms such as Facebook.

© The Author(s) 2017
E. Khoo et al., *Software Literacy*, SpringerBriefs in Education,
https://doi.org/10.1007/978-981-10-7059-4